T0235795

Modeling for Hybrid and Electric Vehicles Using Simscape

Synthesis Lectures on Advances in Automotive Technology

Editor
Amir Khajepour, *University of Waterloo*

The automotive industry has entered a transformational period that will see an unprecedented evolution in the technological capabilities of vehicles. Significant advances in new manufacturing techniques, low-cost sensors, high processing power, and ubiquitous real-time access to information mean that vehicles are rapidly changing and growing in complexity. These new technologies—including the inevitable evolution toward autonomous vehicles—will ultimately deliver substantial benefits to drivers, passengers, and the environment. Synthesis Lectures on Advances in Automotive Technology Series is intended to introduce such new transformational technologies in the automotive industry to its readers.

Deep Learning for Autonomous Vehicle Control: Algorithms, State-of-the-Art, and Future Prospects
Sampo Kuuti, Saber Fallah, Richard Bowden, and Phil Barber
2019

Narrow Tilting Vehicles: Mechanism, Dynamics, and Control
Chen Tang and Amir Khajepour
2019

Dynamic Stability and Control of Tripped and Untripped Vehicle Rollover
Zhilin Jin, Bin Li, and Jungxuan Li
2019

Real-Time Road Profile Identification and Monitoring: Theory and Application
Yechen Qin, Hong Wang, Yanjun Huang, and Xiaolin Tang
2018

Noise and Torsional Vibration Analysis of Hybrid Vehicles
Xiaolin Tang, Yanjun Huang, Hong Wang, and Yechen Qin
2018

Smart Charging and Anti-Idling Systems
Yanjun Huang, Soheil Mohagheghi Fard, Milad Khazraee, Hong Wang, and Amir Khajepour
2018

Design and Avanced Robust Chassis Dynamics Control for X-by-Wire Unmanned Ground Vehicle
Jun Ni, Jibin Hu, and Changle Xiang
2018

Electrification of Heavy-Duty Construction Vehicles
Hong Wang, Yanjun Huang, Amir Khajepour, and Chuan Hu
2017

Vehicle Suspension System Technology and Design
Avesta Goodarzi and Amir Khajepour
2017

Modeling for Hybrid and Electric Vehicles Using Simscape

Shuvra Das

ISBN: 978-3-031-00380-6   paperback
ISBN: 978-3-031-01508-3   ebook
ISBN: 978-3-031-00012-6   hardcover

DOI 10.1007/978-3-031-01508-3

A Publication in the Springer series
SYNTHESIS LECTURES ON ADVANCES IN AUTOMOTIVE TECHNOLOGY

Lecture #14
Series Editor: Amir Khajepour, *University of Waterloo*
Series ISSN
Print 2576-8107   Electronic 2576-8131

Modeling for
Hybrid and Electric Vehicles
Using Simscape

Shuvra Das
University of Detroit Mercy

SYNTHESIS LECTURES ON ADVANCES IN AUTOMOTIVE TECHNOLOGY
#14

ABSTRACT

Automobiles have played an important role in the shaping of the human civilization for over a century and continue to play a crucial role today. The design, construction, and performance of automobiles have evolved over the years. For many years, there has been a strong shift toward electrification of automobiles. It started with the by-wire systems where more efficient electro-mechanical subsystems started replacing purely mechanical devices, e.g., anti-lock brakes, drive-by-wire, and cruise control. Over the last decade, driven by a strong push for fuel efficiency, pollution reduction, and environmental stewardship, electric and hybrid electric vehicles have become quite popular. In fact, almost all the automobile manufacturers have adopted strategies and launched vehicle models that are electric and/or hybrid. With this shift in technology, employers have growing needs for new talent in areas such as energy storage and battery technology, power electronics, electric motor drives, embedded control systems, and integration of multi-disciplinary systems. To support these needs, universities are adjusting their programs to train students in these new areas of expertise.

For electric and hybrid technology to deliver superior performance and efficiency, all subsystems have to work seamlessly and in unison every time and all the time. To ensure this level of precision and reliability, modeling and simulation play crucial roles during the design and development cycle of electric and hybrid vehicles. Simscape, a Matlab/Simulink toolbox for modeling physical systems, is an ideally suited platform for developing and deploying models for systems and sub-systems that are critical for hybrid and electric vehicles. This text will focus on guiding the reader in the development of models for all critical areas of hybrid and electric vehicles. There are numerous texts on electric and hybrid vehicles in the market right now. A majority of these texts focus on the relevant technology and the physics and engineering of their operation. In contrast, this text focuses on the application of some of the theories in developing models of physical systems that are at the core of hybrid and electric vehicles. Simscape is the tool of choice for the development of these models. Relevant background and appropriate theory are referenced and summarized in the context of model development with significantly more emphasis on the model development procedure and obtaining usable and accurate results.

KEYWORDS

electric vehicle, EV, hybrid electric vehicle, HEV, Matlab/SIMULINK, Simscape, multidisciplinary, system modeling, electrification, power electronics, electric drive, battery

Contents

Preface

As per current statistics, the number of motor vehicles in the world is about 1.4 billion. In the United States alone the number is about 284 million and growing. Around 2009, China overtook the U.S. to become the world's largest producer of automobiles. In 2020, about 20 million passenger cars and about five million passenger vehicles were produced in China. An overwhelming majority of vehicles across the world are heavily reliant on fossil fuels. This puts enormous pressure on the limited supply of fossil-fuel resources and more importantly, is having a perceptibly adverse impact on global climate. In humankind's effort to seek solutions to the climate crisis and reduce our dependence on fossil fuel, electrification of automobiles has emerged as a key strategy. Although efforts of manufacturing electric cars go back to the early days of automobile manufacturing, significant development over the last two decades have made Electric Vehicles (EV) and Hybrid Electric Vehicles (HEV) commercially viable. Almost all automobile manufacturers are producing fully electric or hybrid electric versions of their models and companies such as Tesla are devoted to manufacturing *only* electric vehicles for the commercial marketplace.

Design, engineering, and production of EVs and HEVs need a confluence of many disciplines and topic areas such as combustion, mechanical systems, power electronics, embedded control, motor technology and electro-mechanical energy conversion, embedded systems, and software development. Moreover, mathematical modeling and simulation play very critical roles in this process. This book attempts to introduce the reader to system modeling and simulation of EVs and HEVs using Simscape, a toolbox available within Matlab/SIMULINK software suite. Simscape uses the physical network approach which is very intuitive, a technique that will look familiar to users who have had basic courses in engineering. After a brief introduction to the field of EVs and HEVs the rest of the book is devoted to discussing model development in various key aspects of the electrified driveline, e.g., power storage, power electronics, electric drive and motors, and vehicle architecture. The target audience of this book is one who wants to learn how to use modeling and simulation to understand the various aspects of EVs and HEVs, as well as those who would like to develop some basic modeling skills and then use the skill to devise their own solutions to similar problems. After going through this text, the reader should also explore Matlab help files and model libraries which provide a lot of useful and in-depth information on this topic.

Shuvra Das
April 2021

CHAPTER 1

Introduction to Electric Vehicles and Hybrid Electric Vehicles

1.1 INTRODUCTION

Over the last decade and half, excitement and interest in Electric Vehicles (EVs) and Hybrid Electric Vehicles (HEVs) have grown significantly. This interest has been driven by greater awareness of climate change and the role that fossil fuels play in it, significant advancement in battery technology, stricter regulation around emission standards, availability of EV and HEV models without any diminished performance capability, and several other factors. All the automobile companies that have manufactured fossil-fuel-powered vehicles for many decades now have well-established electrification programs and product lines. Moreover, new companies such as Tesla have established significant market presence in this sector.

1.2 BRIEF HISTORY OF EVS AND HEVS

The concept of EVs and HEVs is not new. In fact, in the early days of the automobile electric-powered vehicles showed a lot more promise until the 1910s. The invention of the electric starter and Henry Ford's innovation of the assembly line leading to cheaper internal combustion engine (ICE) cars altered the course in favor of the gasoline-driven vehicles. The following provides some of the important dates and events in history that relate to vehicle electrification.

1.2.1 EARLY YEARS

- Pre-1830: Steam-powered transportation was the order of the day.

- 1831: Faraday's law was formulated and the DC motor was invented.

- 1832–1839: Scottish inventor Robert Anderson invented the first crude electric carriage powered by non-rechargeable primary cells.

- 1835: American Thomas Davenport is credited with building the first practical electric vehicle—a small locomotive.

- 1859: French physicist Gaston Planté invented the rechargeable lead-acid storage battery. In 1881, his countryman Camille Faure improved the storage battery's ability to supply current and invented the basic lead-acid battery used in automobiles.

- 1874: Battery-powered carriage was first built.

- 1891: William Morrison of Des Moines, Iowa builds the first successful electric automobile in the United States.

- 1893: A handful of different makes and models of electric cars are exhibited in Chicago.

- 1897: The first electric taxis hit the streets of New York City early in the year. The Pope Manufacturing Company of Connecticut becomes the first large-scale American electric automobile manufacturer.

- 1899: Believing that electricity will run autos in the future, Thomas Alva Edison begins his mission to create a long-lasting, powerful battery for commercial automobiles. Although his research yields some improvements to the alkaline battery, he ultimately abandons his quest a decade later.

- 1900: The electric automobile is in its heyday. Of the 4,192 cars produced in the United States, 28% are powered by electricity, and electric autos represent about one-third of all cars found on the roads of New York City, Boston, and Chicago.

- 1908: Henry Ford introduces the mass-produced and gasoline-powered Model T, which will have a profound effect on the U.S. automobile market.

- 1911: Invention of starter motor makes gasoline vehicles easier to start (Kettering).

- Improvements in mass production of gas-powered vehicles which sold for $260 in 1925 compared to $850 in 1909. EVs were more expensive.

- Rural areas had very limited access to electricity to charge batteries, whereas gasoline could be sold in those areas. So, presence of gasoline-driven cars in the marketplace grew significantly.

1.2.2   1960S AND 1970S

- Resurgence of EV research and development in the 1960's is due to increase awareness of air quality.

- Congress introduces bills recommending the use of EVs as a means of reducing air pollution.

- Major Internal Combustion Engine Vehicles (ICEV) manufacturers become involved in EV R&D (e.g., GM, Ford).

- In the 1970s, gasoline prices increase dramatically as energy crisis increases. This led to immense interest in EV.

- The Arab oil embargo of 1973 increases demand for alternate energy sources. Less dependence on foreign oil becomes desirable.

- In 1975, 352 electric vans were delivered to the U.S. postal service for testing.

- In 1976, Congress enacts Public Law 94-413, the *Electric and Hybrid Vehicle Research, Development and Demonstration Act of 1976*. This act authorizes a federal program to promote electric and hybrid vehicle technologies and to demonstrate the commercial feasibility of EVs.

1.2.3   1980s AND 1990s

- Development of magnetic bearings used in flywheel energy storage systems.

- Improvements of high-power, high-frequency semiconductor switches, along with μ-processor revolution, led to improved power converter design.

- Legislation passed by the California Air Resources Board: By 1998, 2% of all vehicles (about 40,000) would be zero emission vehicle (ZEV). By 2003, 10% (about 500,000) would be EV. More than one dozen eastern states also adopted this law to comply with federal regulations on emission standards.

- SERA (Solar and Electric Racing Association) organizes electric car races around the country. This organization wants to do for EVs what Indy Racing does for ICEVs (improvements and developments through racing).

- In the 1990s, EVs failed again; possible reasons: range limitation, cheap gasoline, consumers wanted large SUVs and mini-vans, oil companies lobbied hard, CARB switched its mandate at the last minute from electric to hydrogen vehicles, and infrastructure was not well developed.

1.2.4   2007–PRESENT

- Due to the economic downturn government policies were introduced to help the EV and HEV industries invest in them.

- More acceptance from consumers.

- Government subsidy.

- International push for EVs to fight pollution and climate change.

- EVs and HEVs are being accepted by customers. With advancement in battery research and new battery technology, significant range improvements have been achieved.

- Charging infrastructure growth is still taking time.

1.3   EVS AND HEVS

An EV is an electric vehicle that is operated with power from a battery only. In EVs, the sole propulsion is by an electric motor. A traditional vehicle has sole propulsion from an internal combustion engine or a diesel engine. An HEV is a multi-energy source vehicle where, apart from the electric motor, other energy source can be gas, natural gas, battery, ultra-capacitor, fly wheel, solar panel, etc. As per the arrangement of the energy sources, the HEV could be considered a:

- Parallel HEV: where input from multiple sources can be combined, or individual sources can drive the vehicle by itself.

- Series HEV: sole propulsion is by an electric motor, but the electric energy comes from another on board energy source, such as ICE.

- Simple HEV: such as diesel electric locomotive for which energy consumption is not optimized; these are only designed to improve performance (acceleration, etc.).

- Complex HEV: these can possess more than two electric motors, energy consumption and performance are optimized, and have multimode operation capability.

- Heavy hybrids—trucks, locomotives, diesel hybrids, etc.

According to the onboard energy sources, hybrids can be many types, such as: ICE hybrids, diesel hybrids, fuel cell hybrids, solar hybrids (race cars, for example), natural gas hybrids, hybrid locomotive, and heavy hybrids. Currently, the hybrid vehicles that are commercially available are gasoline hybrids where a gasoline engine and a battery-powered motor are the two sources of power. In this text our discussion will remain confined to only this type of hybrid vehicles.

Today's EVs, where electricity is the sole source of power, have some critical drawbacks. They include high initial cost, short driving range, and lengthy recharge time. The range limitation is a significant issue since people need a vehicle not only for commuting (city driving), but also for the long-distance highway driving for vacation and pleasure trips. Recharging takes a much longer time than refueling gasoline and unless infrastructure for instantly replaceable battery cartridges is available (something like home BBQ propane tank replacing), customers may not accept this long recharging delays. Also, the battery pack takes space and weight of the vehicle which otherwise is available to the customer. As a comparison, the key drawbacks of ICE vehicles include high rate of fossil fuel consumption, dependence on foreign oil, high emission, air pollution, global warming and associated environmental hazards, high maintenance cost, and noise.

An HEV can be thought of as an option which carries the best of both worlds. With proper design the fuel economy of HEVs can be optimized in a way that is better than ICE vehicles. This involves running the internal combustion engine always at the optimum efficiency, starting and stopping the ICE as needed, recovering kinetic energy from braking, and using a smaller engine. Also optimized HEVs result in reduced emissions from the ICE. HEVs make a lot less noise than ICEs because of the smaller engine size, non-continuous engine operation, and quiet motor operations. Finally, the ICE related maintenance needs are significantly reduced leading to fewer tune ups, longer life cycle of ICE, fewer spark-plug changes, fewer oil changes, fewer fuel filters, antifreeze, radiator flushes or water pumps, fewer exhaust repairs, or muffler changes.

There have been some concerns about HEVs, especially when they first models started coming to the market. With the inclusion of newer technology such as battery, electric machines, motor controllers, etc., the initial cost of HEVs were higher. But over time they are coming down. With the use of newer technology there were concerns about their reliability in the minds of the customer. And the dealership and repair shops were not familiar with new components and new technology. This includes the safety concerns around working with high-voltage batteries used in HEVs. While these concerns were not trivial, they are definitely transient. With the passage of time most of these go away.

1.4   INTERDISCIPLINARY NATURE OF HEV

The automobile has always been considered as one of the most complex pieces of machine which was primarily made of many mechanical components and subsystems. Over time, with the advent of newer, better, and more efficient technology many mechanical devices were being replaced by electro-mechanical ones. This transition to by-wire systems was also accompanied by ever increasing implementation of software-based controllers through many of the embedded computers that were added in the automobile. The HEV technology has accelerated this transition of an automobile from a purely mechanical device to something that is highly interdisciplinary. Figure 1.1 provides a pictorial overview of some of the main topical area that are critical for the design, engineering, and production of HEVs and EVs. This is a rather long list of topics and the interdisciplinary nature of these topics is also noticeable.

Figure 1.2 shows a Ford Hybrid Escape's powertrain. In the picture, the main subsystems of a Hybrid vehicle are mentioned. The two power sources are the Internal Combustion (IC) engine and the motor-generator pair used for traction, a Nickel-Metal Hydride battery to supply electric power, power electronics, and vehicle controller module shown in the picture as the Vehicle System Controller, and a vehicle system architecture design that helps to support the HEV's response in all types of driving conditions. In our exploration of modeling and simulation,

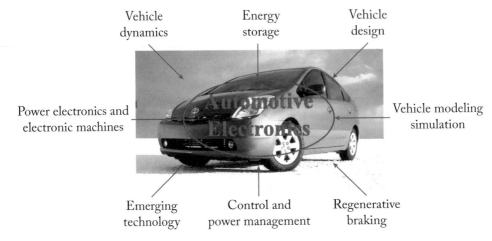

Figure 1.1: Technical topics critical for HEVs show a confluence of many disciplines.

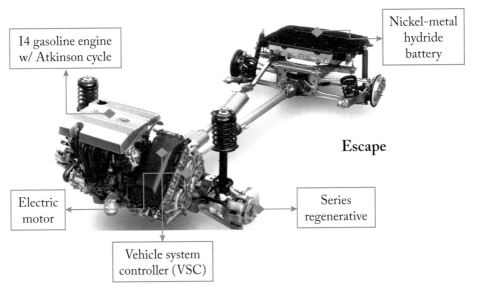

Figure 1.2: Key subsystems relevant to an HEV vehicle.

we will explore all of these different sub-areas, energy storage, motor drives (AC and DC), power electronics, and system architectures.

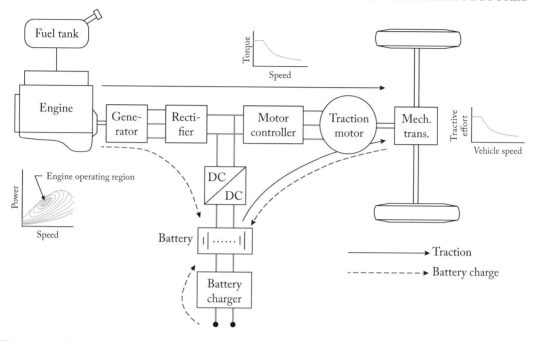

Figure 1.3: Series architecture.

1.5   VEHICLE ARCHITECTURES

HEVs that are available in the market place have demonstrated different vehicle architectures. We will present modeling examples of a few of those in a later chapter. The two main types are series and parallel architectures and we will briefly discuss the key features of both in this section. There are other architectures which are some combination of series and parallel. We are going to discuss all those combinations here.

1.5.1   SERIES ARCHITECTURE

In the series architecture the main source of power is the engine (Figure 1.3). The engine drives a generator to convert mechanical energy to electrical energy. The generator drives a motor which drives the car. The generator is also connected to the battery so that any excess power is used to re-charge the battery. Also, during braking, the recovered energy is used to re-charge the battery. Controllers are designed to ensure that the system performs at the optimum level during all the modes of operation. The different operation modes for a series HEV are as follows.

- Battery alone mode: engine is off, vehicle is powered by the battery only.

- Engine alone mode: power from ICE/generator to motor.

- Combined mode: both ICE/generator set and battery provides power to the traction motor.

- Power split mode: ICE/generator power split to drive the vehicle and charge the battery.

- Stationary charging mode: ICE/G is used to charge the battery.

- Regenerative braking mode: where the recovered energy is used to charge the battery.

Some of the advantages of series architecture include: the engine operation is optimized so that the engine can operate at a setting where the efficiency is the highest and the engine itself can be redesigned, smaller engine is possible since all the excess energy is being harnessed. Also, since the engine can be operated at a particular setting use of a high-speed engine is possible, and a transmission is not really required in this setting. The control algorithm is not too complicated either. Some of the chief disadvantages are: energy is being converted twice, from fuel to ICE/generator to motor and battery, additional weight/cost due to increased components, traction motor, generator, and ICE all have to be full-sized to meet the vehicle performance needs.

1.5.2   PARALLEL ARCHITECTURE

In the parallel architecture there are two main sources of power: the engine and the battery/motor (Figure 1.4). The engine and the motor are mechanically coupled. This coupling arrangement has many possible designs including an additional generator as well. The control algorithm optimizes the efficiency of this system by continuously balancing the power drawn from the two sources in an optimal fashion. During braking, the recovered energy is used to re-charge the battery. The different operation modes for a paralel HEV are as follows.

- Motor alone mode: engine is off, vehicle is powered by the battery/motor only.

- Engine alone mode: ICE drives the vehicle just by itself.

- Combined mode: both ICE and motor provide power to drive the vehicle.

- Power split mode: ICE power split to drive the vehicle and charge the battery.

- Stationary charging mode: ICE is used to charge the battery.

- Regenerative braking mode: where the recovered energy is used to charge the battery.

Some of the advantages of parallel architecture include: the engine operation is optimized so that the engine can operate at a setting or by optimal sharing of power between the engine and the motor, smaller engine is possible since the total power need is shared between multiple sources. Plug-in hybrid is definitely possible which further improves the fuel economy and

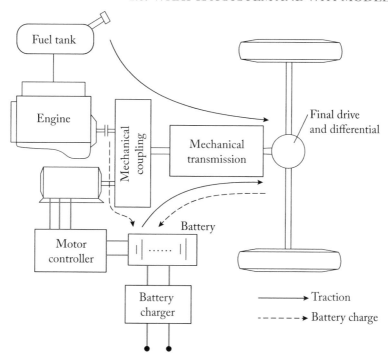

Figure 1.4: Parallel architecture.

emission reduction. Some of the chief disadvantages include the need for a reasonably complex transmission setup to couple the two sources of power in an optimal fashion and complex control algorithms to ensure properly optimized operation in different stages of the drive cycle.

1.6 WHAT IS A SYSTEM AND WHY MODEL SYSTEMS?

Mathematical system models and their solutions become powerful tools in the hands of system designers. They can be used for a variety of purposes such as the following.

- **Analysis:** For given input and known system (and state variables), what would be the output?

- **Identification:** For given input history and known output history, what would the model and its state variables be?

- **Synthesis:** For given input and a desired output, design the system (along with its state variables) so that the system performs the way that is desired.

Learning how to develop useful system models takes time and experience. We therefore go about the above three activities in the order that they are stated. Beginning system modelers spend a lot of time learning to "analyze" systems. Only after a good bit of experience do they venture into the world of "identification" of system. And "Synthesis" requires the maximum amount of experience in the field.

Because a model is somewhat of a simplification of the reality, there is a great deal of art in the construction of models. An overly complex and detailed model may contain parameters virtually impossible to estimate, and may bring in irrelevant details which may not be necessary. Any system designer should have a way to find models of varying complexity so as to find the simplest model capable of answering the questions about the system under study. A system could be broken into many parts depending on the level of complexity one needs. System analysis through a breakdown into its fundamental components is an art in itself and requires some expertise and experience.

In this book we will go through a process of developing system models for EVs and HEVs. We will use appropriate levels of simplification and approximation in the models to ensure that they are informative yet not overly complex for first-time learners. The process of model development will be focused mainly toward the goal of analyzing system behavior. We hope that with some practice in the area of system analysis the students would be ready to start tasks in system identification and design.

1.7   MATHEMATICAL MODELING TECHNIQUES USED IN PRACTICE

Many different approaches have been used in the development of system models. One of the most common methods is deriving the state-space equations from first principles, namely from Newton's laws of mechanics, Kirchoff's voltage and current laws for electrical circuits, power flow analysis, etc. These relationships are then numerically solved to obtain system responses. There are several graphical approaches which are quite popular among different technical communities. Linear graphs is one of them, where state-space equations are modeled as block diagrams connected by paths showing the flow of information from one block to another. Figure 1.5 shows a SIMULINK model of a permanent magnet DC motor built by joining different SIMULINK function blocks with proper information flow paths. This block diagram is based on signal flow, i.e., only one piece of information flows through the connections, and it is a signal rather than a physical quantity with specific meaning. The SIMULINK model is essentially a representation of the actual mathematical equations that govern the behavior of the system and in order to build such a model one has to know or derive the equations. One of the biggest challenges in building a representative system model is user-knowledge of the system. Someone who knows a system well enough including all the mathematical equations may find it easy to replicate the equations in a tool such as SIMULINK. However, someone who is encountering a system the first time may find it quite difficult to transition from a physical world to the level of abstraction necessary for a mathematical model.

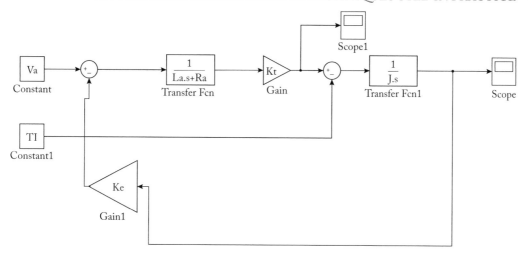

Figure 1.5: A signal flow diagram of a permanent magnet DC motor modeled in SIMULINK.

An important step in all of these methods can be the derivation of the governing relation-ships. Within a single domain (Mechanical, Electrical, etc.) deriving the governing equations may not be difficult because we may be within our specific area of expertise but when we work in a multi-domain environment it becomes somewhat more difficult for someone who is not suitably trained. As was briefly mentioned earlier, the EV and HEV technology is a confluence of many disciplines and sub-disciplines. Within each discipline of engineering education, system repre-sentation and solution techniques have evolved along different paths. We are trained to think in terms of statics, dynamics, circuit analysis, electromagnetism, hydraulics, etc. to be different subject areas where different solution techniques are used for problem solving. These artificial barriers between different disciplines or subjects highlight the differences without providing a hint of the fact that the underlying similarities are much more than the perceived differences.

There have been other tools developed which use a different approach known as phys-ical object-based model. In this approach components or subsystems are available as icons or blocks. Component behavior or its constitutive model is already programmed as part of the icon/block model. Thus, a capacitor block, mass block, or battery block looks similar to how they are represented in technical literature and models the constitutive equations of those spe-cific components. These components are joined to each other to model the system much the same way a signal diagram is drawn. However, all connections carry power information rather than signal information. For a mechanical system, power is a product of force and velocity, whereas power in an electrical system is voltage times current. The connections essentially are connecting components that exchange two types of physical quantities that make up power and these components vary from one domain to another. This is quite different from a signal flow diagram like in a traditional SIMULINK model. Thus, a physical model looks quite similar to

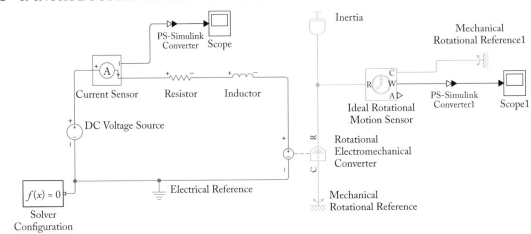

Figure 1.6: A motor model in Simscape.

the schematic diagrams that are drawn in literature to represent such physical systems and are therefore more familiar to the user than a signal flow diagram. Finally, the model developer need not derive the governing differential equations before putting a successful model together. This is a huge advantage since the important step of mathematical abstraction can be bypassed and the model developer can be productive with confidence in a hurry.

Simscape is a Matlab tool available within the Simulink suite that uses a physical modeling approach for the development of System models. It employs the Physical Network approach, which differs from the standard Simulink modeling approach and is particularly suited to simulating systems that consist of real physical components. Figure 1.6 shows a Simscape model of the DC motor, where an inductor or an inertia or a resistor looks like what they represent and the connections exchange physically meaningful quantities such as voltage, current, torque, etc.

1.8   SOFTWARE

In this text we have used Simscape for all the model development discussion and simulation results that are reported. As mentioned before, Simscape is a Matlab/SIMULINK toolbox and is available with a standard license of Matlab. The text is divided into logical sections to cover all aspects of EV and HEV systems. Effort was made to ensure that a first-time user of the tool can start with this book and be productive fairly quickly. Matlab/Simulink is a very well-established tool and has extensive on-line documentation libraries. The user will be able to answer a lot of their own questions beyond what is discussed here by referring to the Matlab libraries.

CHAPTER 2

Introduction to Simscape and Vehicle Dynamics

2.1 PHYSICAL NETWORK APPROACH TO MODELING USING SIMSCAPE

Simscape is a toolbox available within the Simulink environment. It consists of a set of block libraries and simulation features for modeling physical systems. Simscape is based on the Physical Network approach, which differs from the standard Simulink modeling approach and is particularly suited to simulating systems that consist of real physical components.

In Simulink models, mathematical relationships are modeled using blocks connected by signal carrying links. The blocks perform specific mathematical functions. Simscape uses blocks that represent physical objects and these blocks replicate the constitutive relations that govern the behavior of the physical object represented by the block. A Simscape model is a network representation of the system under design, based on the Physical Network approach. Most often the network resembles typical schematics of the systems that the user is familiar with. Thus, an electric or magnetic circuit looks like a circuit seen in texts. The same is true for other domains. The exchange currency of these networks is power (or energy flow over time), i.e., the elements transmit power among each other through their points of entry or exit which are also known as ports. These connection ports are nondirectional. They are similar to physical connections between elements. Connecting Simscape blocks together is analogous to connecting real components, such as mass, spring, etc. Simscape diagrams essentially mimic the physical system layout. Just like real systems, flow directions need not be specified when connecting Simscape blocks. In this physical network approach the two variables that make up power are known as the Through Variable (TV) and Across Variable (AV).

The number of connection ports for each element is determined by the number of energy flows it exchanges with other elements in the system or the number of points through which power may enter or leave the element. For example, a permanent magnet DC motor is a two-port device with electric power coming in from one port and mechanical or rotational power leaving from the other port.

Energy flow or power is characterized by two variables. In a Simscape modeling world, these two variables are known as TV and AV. Usually, these are the variables whose product is the energy flow in watts and can be considered the basic variables. For example, the basic

Table 2.1: Through and Across variables in different domains

Physical Domain	Across Variable	Through Variable
Electrical	Voltage	Current
Hydraulic	Pressure	Flow rate
Magnetic	Magnetomotive force (mmf)	Flux
Mechanical rotational	Angular velocity	Torque
Mechanical translational	Translational velocity	Force
Thermal	Temperature	Heat flow

variables for mechanical translational systems are force and velocity, for mechanical rotational systems—torque and angular velocity, for hydraulic systems—flow rate and pressure, for electrical systems—current and voltage.

2.1.1   VARIABLE TYPES

The TVs and AVs variables for different domains are listed in the Table 2.1. TVs are measured with a gauge connected in series to an element and AVs are measured with a gauge connected in parallel to an element.

Generally, the product of each pair of Across and Through variables associated with a domain is power (energy flow in watts). The exceptions are magnetic domain (where the product of mmf and flux is not power, but energy).

2.1.2   DIRECTION OF VARIABLES

The variables used in calculation are characterized by their magnitude and sign. The sign is the result of measurement orientation. The same variable can be positive or negative, depending on the polarity of a measurement gage. Simscape library has sensor elements for each domain to measure variables.

Elements with only two ports are characterized with one pair of variables, a TV and an AV. Since these variables are closely related, their orientation is defined with one direction. For example, if an element is oriented from port A to port B (Figure 2.1), it implies that the TV is positive if it "flows" from A to B, and the AV is determined as $AV = AV_A - AV_B$, where AV_A and AV_B are the element node potentials or, in other words, the values of this AV at ports A and B, respectively.

This approach to the direction of variables has the following benefits.

- Provides a simple and consistent way to determine whether an element is active or passive. Energy is one of the most important characteristics to be determined during simulation. If the variables direction, or sign, is determined as described above, their

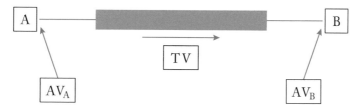

Figure 2.1: A generic element oriented from port A to port B.

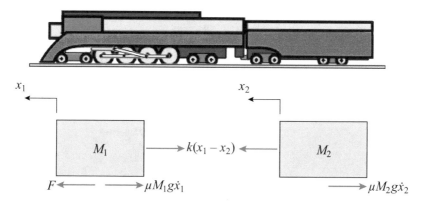

Figure 2.2: Two-Mass train problem.

product (that is, the energy flow) is positive if the element consumes energy, and is negative if it provides energy to a system. This rule is followed throughout the Simscape software.

- Simplifies the model description. Symbol $A \rightarrow B$ is enough to specify variable polarity for both the Across and the Through variables.

The following rudimentary example illustrates a Physical Network representation of a simple two-mass problem. In this example, we consider a toy train consisting of an engine and a car. Assuming that the train only travels in one dimension (along the track), the goal is to control the train so that it starts and stops smoothly, and so that it can track a constant speed command with minimal error in steady state (Figure 2.2).

The mass of the engine and the car are represented by M_1 and M_2, respectively. Furthermore, the engine and car are connected via a coupling with stiffness k. The coupling is modeled as a spring with a spring constant k. The force F represents the force generated between the wheels of the engine and the track, while μ represents the coefficient of rolling friction.

The accompanying figure shows the free body diagram of the system with the different forces acting on it. Using Newton's law and summation of forces for the two masses the governing

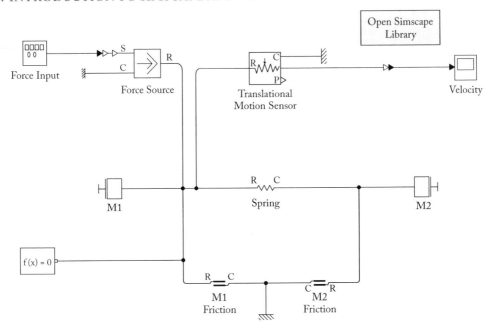

Figure 2.3: Simscape model of the two-mass train problem.

equations may be written as:

$$M_1\ddot{x}_1 = F - k(x_1 - x_2) - \mu M_1 g \dot{x}_1$$

$$M_2\ddot{x}_2 = k(x_1 - x_2) - \mu M_2 g \dot{x}_2. \tag{2.1}$$

The corresponding Simscape model for the system is shown in Figure 2.3.

This model is a fairly simplified version of the actual problem. However, the advantage of the networked system is that it is easily scalable; for example, a linear spring can be replaced by a nonlinear spring model or by a spring damper combination, if needed, without affecting any other parts of the model. The simple friction element can be swapped with a tire-road model if, for example, the two masses were representing a truck-trailer combination on a road rather than a train.

All the elements in a network are divided into active and passive elements, depending on whether they deliver energy to the system or dissipate/store it. Active elements (force and velocity sources, flow rate, and pressure sources, etc.) must be oriented strictly in accordance with the line of action or function that they are expected to perform in the system, while passive elements (dampers, inductors, resistors, springs, pipelines, etc.) can be oriented either way.

In the two-mass train model the different elements, masses, the spring, and the friction elements all are shown to have two ports, R and C. All these are passive elements. The R and C

port connections of the passive elements could be reversed in the network without affecting any calculated results. For the force source, an active element, the block positive direction is from port C to port R. In this block, port C is associated with the source reference point (ground), and port R is associated with the $M1$. This means the force is positive if it acts in the direction from C to R, and causes bodies connected to port R to accelerate in the positive direction. The relative velocity is determined as $v = vC - vR$, where vR, vC are the absolute velocities at ports R and C, respectively, and it is negative if velocity at port R is greater than that at port C. The power generated by the source is computed as the product of force and velocity, and is positive if the source provides energy to the system. All this means that if the connections are reversed in the force source or any other similar active elements, an opposite effect will be observed.

For more information on this check the block source or the block reference page on Simscape help pages if in doubt about the block orientation and direction of variables.

2.1.3   ELEMENT TYPES

All systems are made of a few basic components or elements that can be categorized into a few broad categories based on how they behave. It is important to understand these basic behaviors of elements to realize that there are plenty of similarities between systems in different domains. Within Simscape the element types are broadly categorized as Passive Elements and Active Elements. In a network all elements are either active or passive elements, depending on whether they deliver energy to the system or dissipate (or store) it.

2.1.4   PASSIVE ELEMENTS

Among the many different passive elements that are available, element behavior can determine how they can be categorized. There are energy storage elements such as the capacitor or spring, the inductor or mass/inertia, and dissipative elements such as damper and resistor. There are also energy transfer elements such as transformers, gear trains, levers, etc. and gyrator elements such as electric motors of various kinds, solenoid coils, and other similar devices.

2.1.5   ACTIVE ELEMENTS

There are in general two types of sources that are used in these types of system models. Both these sources supply energy (or power) but in one case the power is supplied with a known TV and in the other case the power is supplied with a known AV. For example, a voltage source in an electrical system supplies power to an electrical circuit using a defined or known voltage profile. The current drawn is determined by the load of the system that is receiving the power. Similarly, a velocity source in a mechanical system supplies power with a known velocity profile and the force part of the power is determined by the system that is receiving power. All these are active elements and the direction of the voltage or velocity or force or any other quantity will be reversed with the reversing of polarity of these elements.

2.1.6   CONNECTOR PORTS AND CONNECTION LINES

Simscape blocks may have two types of ports: physical conserving ports and physical signal ports. They function differently and are connected in different ways. These ports and connections between them are described in detail below.

2.1.7   PHYSICAL CONSERVING PORTS

Simscape blocks have special conserving ports. Conserving ports are connected with physical connection lines, distinct from normal Simulink lines. Physical connection lines have no directionality and represent the exchange of energy flows/power, according to the Physical Network approach. Conserving ports can only be connected to other conserving ports of the same type (i.e., mechanical elements are connected by mechanical line types and electrical elements are connected by electrical line type, etc.). The physical connection lines that connect conserving ports together are nondirectional lines that carry physical variables (Across and Through variables) rather than signals. In passive and active element blocks in Simscape they are shown at C and R. Physical connection lines cannot be connected to Simulink ports or physical signal ports. In the example exercise shown in Figure 2.3 the connection lines connecting the masses with the spring and the friction elements are all Physical Conserving ports.

Two directly connected conserving ports must have the same values for all their Across variables (such as pressure or angular velocity). Branches can be added to physical connection lines. When branching happens, components directly connected with one another continue to share the same Across variables. Any Through variable (such as flow rate or torque) transferred along the physical connection line is divided among the components connected by the branches. This division is determined automatically by the dynamics of the system in consideration. For each Through variable, the sum of all its values flowing into a branch point equals the sum of all its values flowing out. This is similar to the Kirchoff's current law. For improved readability of block diagrams, each Simscape domain uses a distinct default color and line style for the connection lines. For more information, see Domain-Specific Line Styles in Simscape Help.

2.1.8   PHYSICAL SIGNAL PORTS

Physical signal ports carry signals between Simscape blocks. They are connected with regular connection lines, similar to Simulink signal connections. Physical signal ports are used in Simscape block diagrams instead of Simulink input and output ports to increase computation speed and avoid issues with algebraic loops. Physical signals can have units associated with them. The units along with the parameter values can be specified in the block dialogs, and Simscape software performs the necessary unit conversion operations when solving a physical network. Simscape Foundation library contains, among other sublibraries, a Physical Signals block library. These blocks perform math operations and other functions on physical signals, and allow you to graphically implement equations inside the Physical Network. Physical signal lines also have a distinct style and color in block diagrams, similar to physical connection lines in Simulink. In

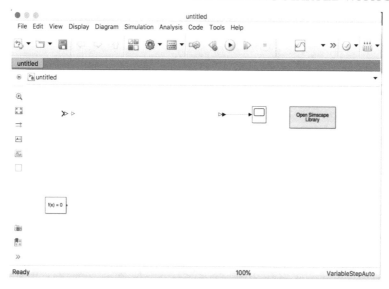

Figure 2.4: A new model file.

the model shown in Figure 2.3 the connection between the Force Input block to the S port of the force actuator is a physical signal connector. Similarly, the output from the velocity sensor that is connected to the V terminal (which is a physical signal port) is a physical signal connection. In both cases the two double arrows are used to transform a Simulink signal to a physical signal. These double arrow symbols represent the **S_PS** (Signal to Physical Signal) or **PS_S** (Physical signal to Signal) transformation.

2.2   GETTING STARTED WITH SIMSCAPE

In the next few chapters many Simscape models and model development will be discussed with appropriate background information. This section will just touch on how to get the model building process started. To open a new Simscape model type "**ssc_new**" in the MATLAB command window. That opens up a new model file and it looks like Figure 2.4.

The file opens with a few basic blocks already included. These are most common blocks that are used in any model. The **Scope** block would be fairly well known to Simulink users. This is used to graphically display any function (e.g., inputs or outputs from a simulation). The two sets of double arrows are the **PS_Simulink** and **Simulink_PS** blocks. These are the links between the traditional Simulink world where we deal with signals (with no physical meaning) and the physical modeling world of Simscape where elements exchange physical variables. The Simulink_PS block changes a Simulink signal to a physical signal whose type and unit can be set by double clicking on the block. And the PS_Simulink block does the exact opposite, i.e., changes a physical signal such as velocity, voltage, displacement, etc. into a signal that can then

Figure 2.5: Simscape library.

be displayed by the scope. The block which shows $f(x) = 0$ is known as the solver configuration block. This can be used to setup a local solver or a global solver. For every model, this block has to be attached some place in the model in order to carry out the simulation.

Once Simscape is invoked (Figure 2.5) one can look for and find a large library of elements in the Simscape library. The Simscape library has several subfolders. The chief among them that we will use in this task are the Foundation Library, the Driveline, and the Electrical library. Figures 2.6, 2.7, and 2.8 show the main screens of these three subfolders. As is obvious from the figures, these subfolders contain many additional subfolders that contain individual elements. It is not possible to go through all of them in detail in this text. A new user of Simscape is strongly encouraged to explore these libraries to get a sense of the large size and diverse content of these libraries. When looking for a specific element to include in a model, trying to locate the element in the library by searching through the library subfolders is not the best approach. It is best to use the following set of tips.

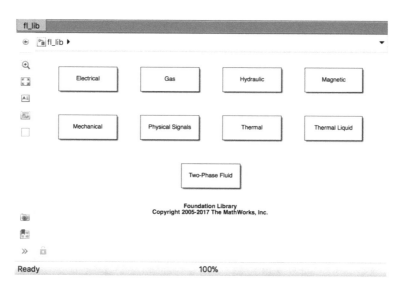

Figure 2.6: Available libraries within Simscape's foundation library.

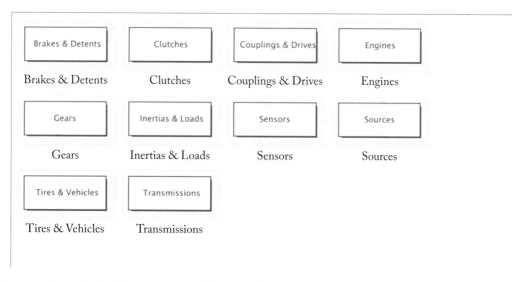

Figure 2.7: Available libraries within Simscape's driveline library.

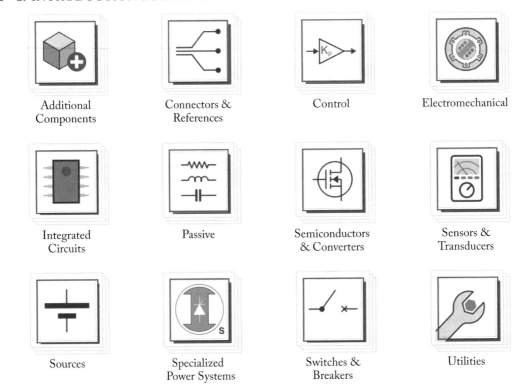

Figure 2.8: Available libraries within Simscape's electrical library.

Tips for adding model elements

1. Use Quick Insert to add the blocks. Click in the diagram and type the name of the block. A list of blocks will appear and you can select the block you want from the list. Alternatively, the Open Simscape Library block can be used to look though the library of all blocks and pick the appropriate one.

2. After the block is entered, a prompt will appear for you to enter the parameter. Enter the variable names as shown below.

3. To rotate a block or flip blocks, right-click on the block and select Flip Block or Rotate block from the Rotate and Flip menu.

4. To show the parameter below the block name, see Set Block Annotation Properties in the documentation.

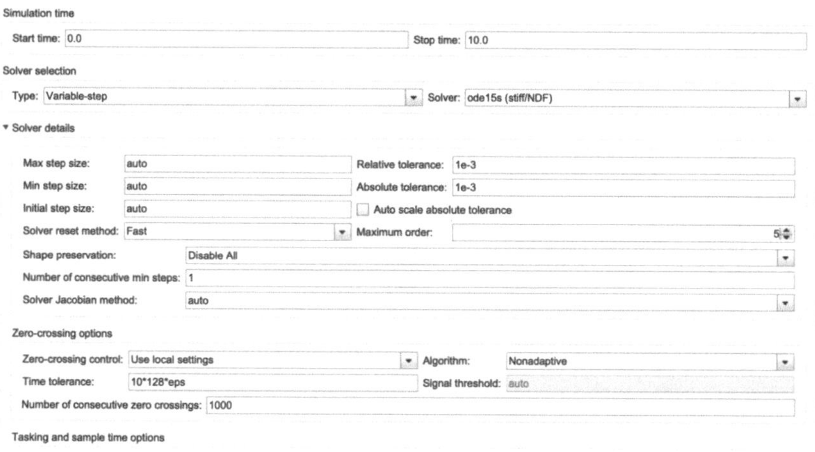

Figure 2.9: Solver setting.

A Note About Solvers: The default solver is **ode 45**. Although this will work fine for most of the examples discussed here it is strongly recommended that you change the solver to a stiff solver (ode15s, ode23t, or ode14x). Do this by selecting "**Configuration Parameters**" from the Simulation menu, selecting the solver pane from the list on the left, and changing the "Solver" parameter to ode15s. Then click OK (Figure 2.9).

The use and need for all these blocks will be better clarified in the next few chapters in the context of the examples that are discussed.

2.3   EXAMPLES

2.3.1   EXAMPLE 2.1: SIMULATION OF VEHICLE LOAD

Here we introduce a fairly simple Simscape model to highlight the vehicle dynamic load that needs to be overcome during driving. The example chosen is a rudimentary model for a vehicle cruise control device. The net traction force F (as shown in Equation (2.2)) applied from the motor drive is used to overcome the three sources of resistance: the road frictional resistance, gravitational resistance if the car is moving up a slope, and the wind resistance due to wind drag. The net force is then utilized to accelerate the vehicle.

$$ma = F - \mu mg - mgSin\theta - \frac{1}{2}\rho C_d A v^2. \tag{2.2}$$

Add a subsystem to the blank model and name it, Vehicle Body. Open the subsystem and add three more subsystems in it, named Gravity, Road, and Drag. Also add, a liner motion sensor, a mass, a mechanical translational reference, a force source, two PS Add blocks, and two PS-S blocks. Also add an Inport, two Outports, and a PMC_port Block.

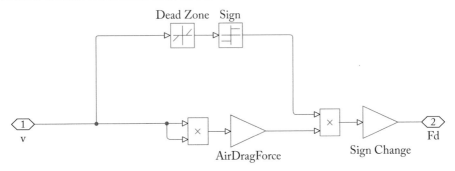

Figure 2.10: Drag subsystem.

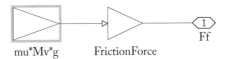

Figure 2.11: Road subsystem.

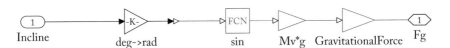

Figure 2.12: Gravity subsystem.

Figures 2.10, 2.11, and 2.12 show the three subsystems that model the three resistance forces from wind drag, road resistance, and gravity, respectively. The mass of the vehicle used is 1200 kg, the coefficient of rolling friction as 0.01, drag coefficent of 0.3, vehicle frontal area of 1.4 sq.m, and air density of 1.225 kg/m^3. The slope angle is accepted as an input in degrees through the inport. All three forces are computed in their respective subsystem and multiplied with -1 and then added. Figure 2.13 shows the entire Vehicle Body subsystem. The net resistive force is then applied using a force source to the Vehicle mass. It also receives input from outside this subsystem in the form of the drive force. The motion sensor tracks the velocity. Double click on the PS-S blocks connected to the motion sensor and choose m/s for the velocity unit and m for the displacement unit and connect them to the two outports.

The cruise control model uses the vehicle model as a subsystem and links it with a control loop. The control loop uses a PI controller to control the vehicle speed to match a desired speed. Double click on the PI controller and set the kp value to be 4400 and Ki to 0.1. Figure 2.14 shows the entire model. A switch block and a step function for force is included in the model for operation without the control loop but for our discussion here the model is always operated with the control loop. In this model we chose not to explicitly model the vehicle power source

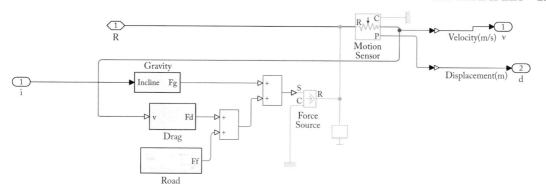

Figure 2.13: Vehicle body subsystem.

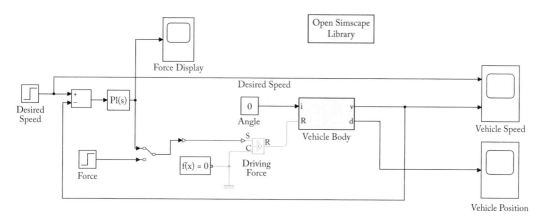

Figure 2.14: Cruise control model.

(engine, battery, etc). Instead, the thrust from the engine is simulated through a control loop and a force actuator that produces the driving force. Later in the book when we specifically explore the different power sources to drive a vehicle such an engine or a motor or both, the vehicle body subsystem that is used here will be reused to model the dynamic load for the vehicle.

The model is simulated for 10 s for two different conditions, a zero degree slope and a 5° slope. The plots for speed comparison, distance traveled, and driving force for the 0° slope are shown in Figures 2.15–2.17. The same three plots for a 5° slope are shown in Figures 2.18–2.20.

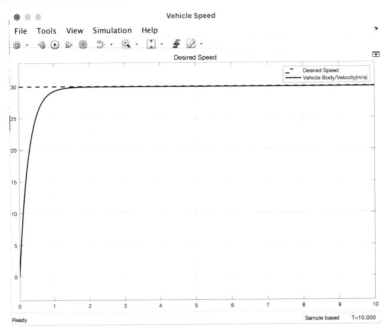

Figure 2.15: Comparison of desired speed and actual speed for driving on a flat surface.

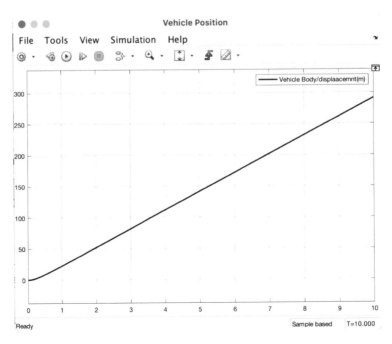

Figure 2.16: Total distance traveled when driving on a flat surface.

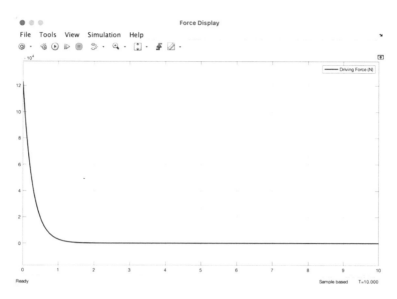

Figure 2.17: Driving force required when driving on a flat surface.

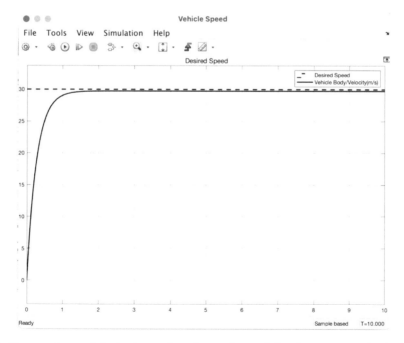

Figure 2.18: Comparison of desired speed and actual speed for driving on a 5° slope.

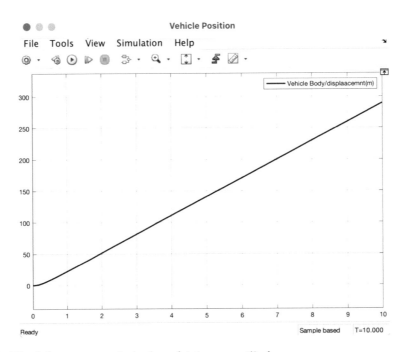

Figure 2.19: Total distance traveled when driving on a 5° slope.

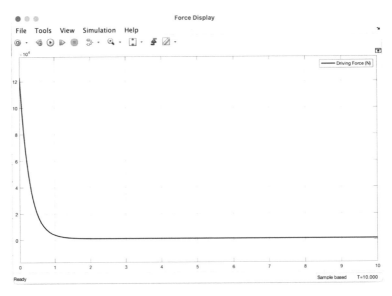

Figure 2.20: Driving force required when driving on a 5° slope.

2.4 SUMMARY

In this chapter we briefly introduced some of the capabilities of Simscape, a Matlab toolbox, along with a quick presentation of a model to replicate the driving load and driving thrust forces. The model was powered from an ideal force source. In future examples we will use a similar vehicle model to test other key subsystems of an EV and HEV.

CHAPTER 3

Modeling Energy Storage

3.1   INTRODUCTION

One of the critical tasks in an EV or HEV is storage of energy, specifically electrical energy. Lead acid batteries have been used in traditional automobiles for a very long time to aid with starting the vehicle and running various devices such as windshield wiper, automated window, etc. They are not quite sufficient for energy storage in EVs and HEVs. Some of the batteries that have emerged as good candidates for electric vehicles are Nickel-Metal Hydride batteries and Lithium-Ion batteries. Along with batteries, another storage device that has found use in these vehicles are ultracapacitors. Storing and releasing energy in batteries is dictated by electro-chemistry while the same phenomenon in capacitors happens through electrostatics. The former is a slow process while the latter process is fast. This difference in time constant of electro-chemical operations and electro-statical operations result in fast charging and discharging capability of capacitors while the same thing happens slowly in the case of batteries. This means that through proper planning and design, these two storage devices can be used in different regimes of operation of electric and hybrid electric vehicles. In stop-and-go city traffic, where kinetic energy from braking can be recovered effectively, the ultracapacitor is very useful because of its faster charging and discharging capability. In highway driving, though, the battery is a better option. In this chapter we look at the battery and ultracapacitor models and explore how they are implemented through appropriate examples.

3.2   EXAMPLES

3.2.1   EXAMPLE 3.1: BATTERY MODEL

Applications[1] associated with the design and operation of EVs and HEVs are almost all multi-disciplinary in nature. There are five broad topics that are key areas of interest in electrified vehicles: (1) power generation and distribution; (2) energy storage; (3) power electronics and electric drives; (4) control algorithms and controller design; and (5) vehicle dynamics. In this chapter we discuss storage of power or energy.

Battery technology and battery research has become a key focal area in vehicle electrification. It is generally agreed that increasing storage capacity and therefore vehicle range (i.e., distance traveled between charges) is necessary to get customers to accept electric vehicles in

[1]Note: This example was also used in the text *Modeling and Simulation of Mechatronic Systems using Simscape* by this author and published by Morgan & Claypool.

large numbers. The first example relates to modeling battery behavior. Even though for many modeling exercises we have assumed batteries or sources of power to be providing a constant voltage, real batteries are not quite like that. As more current is drawn from the batteries the state of charge depletes and along with that, the potential difference or voltage that the battery can supply. For a realistic system model that has to predict longer term operation, a realistic battery model is essential.

A battery block available in the electric sub-directory of Simscape represents a simple battery model. The block has four modeling variants, accessible by right-clicking the block in the block diagram and then selecting the appropriate option from the context menu, under Simscape > Block choices:

Uninstrumented | No thermal port—Basic model that does not output battery charge level or simulate thermal effects. This modeling variant is the default.

Uninstrumented | Show thermal port—Model with exposed thermal port. This model does not measure internal charge level of the battery but is able to model temperature effect.

Instrumented | No thermal port—Model with exposed charge output port. This model does not simulate thermal effects.

Instrumented | Show thermal port—Model that lets you measure internal charge level of the battery and simulate thermal effects. Both the thermal port and the charge output port are available.

The instrumented versions of the battery model have an extra physical signal port that outputs the internal state of charge. This functionality is used to change load behavior as a function of state of charge, without having to build a separate charge state model.

The thermal port exposes a thermal port, which represents the battery thermal mass. When this option is selected, additional parameters need to define battery behavior at a second temperature.

The battery model uses a battery equivalent circuit that is made of the fundamental battery model, the self-discharge resistance (R_{SD}), the charge dynamics model, and the series resistance R_0. Figure 3.1 shows the battery model and the charge dynamics portion explicitly shown. The charge dynamics part uses a series of capacitor-resistor (RC) pairs with different time constants to model dynamic behavior.

The Battery charge capacity parameter can be set to infinite when the block models the battery as a series resistor and a constant voltage source. If Battery charge capacity is set to Finite the block models the battery as charge-dependent voltage source and a series resistor. In the finite case, the voltage is a function of charge and has the following relationship:

$$V = V_0 \left(\frac{SOC}{1 - \beta \, (1 - SOC)} \right), \tag{3.1}$$

where:

SOC (state-of-charge) is the ratio of current charge to rated battery capacity.

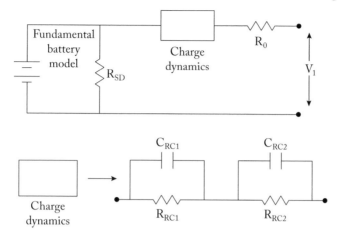

Figure 3.1: Simscape's basic battery model.

V_0 is the voltage when the battery is fully charged at no load, as defined by the Nominal voltage, parameter.

β is a constant that is calculated so that the battery voltage is $V1$ when the charge is $AH1$. The voltage $V1$ and ampere-hour rating $AH1$ are specified using block parameters. $AH1$ is the charge when the no-load (open-circuit) voltage is $V1$, and $V1$ is usually less than the nominal voltage.

3.2.2   CHARGE DYNAMICS

You can model battery charge dynamics using the charge dynamics parameter.

- No dynamics—The equivalent circuit contains no parallel RC sections. There is no delay between terminal voltage and internal charging voltage of the battery.

- The Charge dynamics for the battery can be modeled using a series of parallel RC sections. They are denoted in the setup as First, Second, Third, etc. Time constants depend on how many such RC sections are included in the model. The time constants of each of these sections have to be specified in the block parameters menu.

- Five time-constant dynamics—The equivalent circuit contains five parallel RC sections. Specify the time constants using the First time constant, Second time constant, Third time constant, Fourth time constant, an Fifth time constan parameters

- R_0 is the series resistance. This value is the Internal resistance parameter.

- R_{RC1} and R_{RC2}, etc. are the parallel RC resistances. Specify these values with the First polarization resistance and Second polarization resistance parameters, respectively.

- C_{RC1} and C_{RC2} are the parallel RC capacitances. The time constant τ for each parallel section relates the R and C values using the relationship $C = \tau/R$. Specify τ for each section using the First time constant and Second time constant parameters, respectively.

3.2.3 MODELING THERMAL EFFECTS

For thermal variants of the block, you provide additional parameters to define battery behavior at a second temperature. The extended equations for the voltage when the thermal port is exposed are:

$$V = V_{0T} \left(\frac{SOC}{1 - \beta \left(1 - SOC\right)} \right) \tag{3.2}$$

$$V_{0T} = V_0 \left(1 + \lambda \left(T - T_1\right)\right), \tag{3.3}$$

where:

T is the battery temperature.

T_1 is the nominal measurement temperature.

λ_V is the parameter temperature dependence coefficient for V_0.

β is calculated in the same way as described *before*, using the temperature-modified nominal voltage V_{0T}.

The internal series resistance, self-discharge resistance, and any charge-dynamic resistances are also functions of temperature:

$$R_T = R \left(1 + \lambda_R \left(T - T_1\right)\right), \tag{3.4}$$

where λ_R is the parameter temperature dependence coefficient.

All the temperature dependence coefficients are determined from the corresponding values you provide at the nominal and second measurement temperatures. If you include charge dynamics in the model, the time constants vary with temperature in the same way.

The battery temperature is determined from a simple summation of all the ohmic losses included in the model:

$$M_{th} \dot{T} = \sum_i \frac{V_{T,i}^2}{R_{T,i}}, \tag{3.5}$$

where:

M_{th} is the battery thermal mass (mass * specific heat capacity).

i corresponds to the ith ohmic loss contributor. Depending on how the model is configured the losses include: Series resistance, Self-discharge resistance, First charge dynamics segment, Second charge dynamics segment, Third charge dynamics segment, Fourth charge dynamics segment, and Fifth charge dynamics segment.

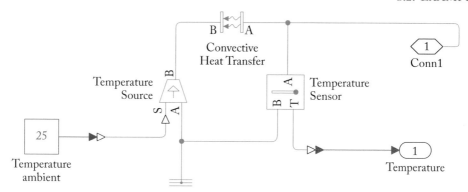

Figure 3.2: Thermal subsystem.

$V_{T,i}$ is the voltage drop across resistor i.

$R_{T,i}$ is resistor i.

Follow the following steps to develop the battery model. (This example was developed using an example in the Simscape help manuals.)

1. Type ssc_new in the Matlab command window to open a new model file.

2. Select and place a Constant Block, a temperature source block, a temperature measurement sensor, and a Convective Heat transfer block from the Heat transfer sub folder in Simscape. Add a Conn1 block and a Thermal reference block. Connect them as shown in Figure 3.2. The battery is the source of heat in this model and the heat generated will be transferred by convection to the surroundings. Set the surrounding temperature to be 25°C. In the convective Heat transfer block set the area to be 0.1 m² and the convective heat transfer coefficient to be 0.1 W/m²K. Now select all the items in this thermal model and make a subsystem by right-clicking and choosing from the menu. Call this subsystem Thermal Aspect.

3. Choose a Battery block from the Electrical folder in Simscape under Sources. Right-click on the block and choose Simscape => Block Choices => Instrumented | Show thermal port.

4. The Battery block settings is a complex set of tables because of the many options that are available. For this model the choices made are shown in Figure 3.3. Each pane is for a particular part of the setting.

5. Add a Voltage Sensor, a Current sensor, and a Current source. Instead of having a load circuit in the model we will use a current source to drive a current load in the circuit to explore its effect on the battery.

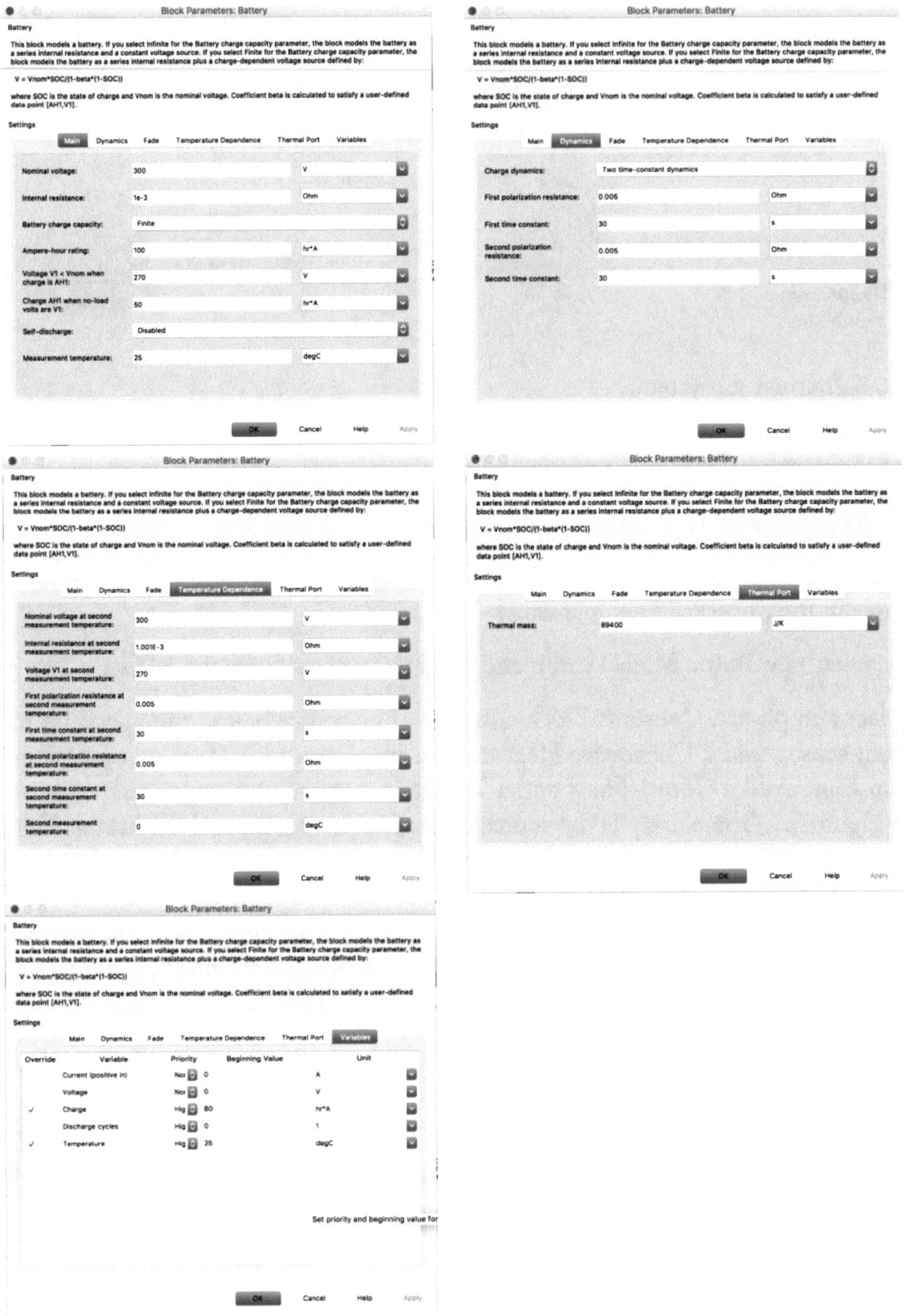

Figure 3.3: Battery model setup.

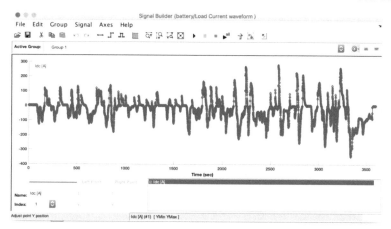

Figure 3.4: Current profile.

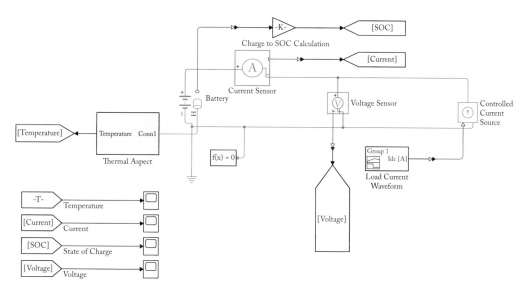

Figure 3.5: Battery model in a circuit.

6. Add a Signal generator and rename it Load Current Waveform. Figure 3.4 shows the waveform that is used. This waveform is similar to a waveform that might be seen in the current drawn in an electric vehicle.

7. Add four From block and four Goto blocks to the model. Create four pairs of From-Goto blocks by naming them Temperature, Voltage, Current, and SOC (state of charge).

8. Attach all the elements as shown in Figure 3.5.

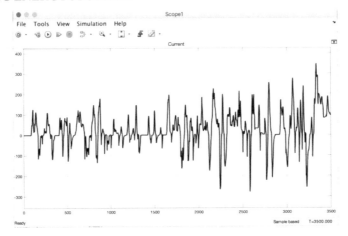

Figure 3.6: Current profile or current load in the circuit.

9. The battery has a terminal (q) which provides data on the amount of charge left in the battery. The SOC (state of charge) is calculated using this data and with a multiplier/gain block. The gain factor is (1/(100*3600)). The 3600 is to convert hours to seconds and 100 is the capacity of battery in hrA.

10. Set the solver setting to Auto and run the simulation for 3500 s.

Figures 3.6, 3.7, 3.8, and 3.9 show the Current profile, Current Temperature, State of Charge, and the Voltage across the battery, respectively, for this model. The SOC plot show how the state of charge drops from initial value of 0.8–0.55 over this length of time. The temperature starts at 25°C and goes up to 26.2°C.

3.2.4   EXAMPLE 3.2: ULTRACAPACITORS

Within Simscape a model is provided for the ultracapacitor (Simscape calls it supercapacitor). The Supercapacitor block represents an electrochemical double-layer capacitor (ELDC). The capacitance values for supercapacitors are many orders of magnitude larger than the values for regular capacitors. Supercapacitors can provide bursts of energy because they can charge and discharge rapidly.

Using the block any number of supercapacitor cells connected in series or in parallel can be modeled. To do so, the relevant parameter, such as Number of series cells or Number of parallel cells, etc., have to be set at the appropriate value. Internally, the block simulates only the equations for a single supercapacitor cell, but it calculates the output voltage and current according to the number of series-connected cells and parallel-connected cells, respectively.

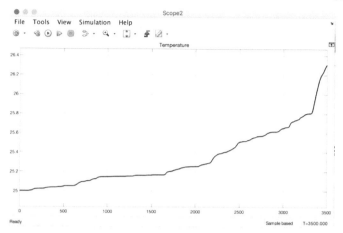

Figure 3.7: Temperature in the battery.

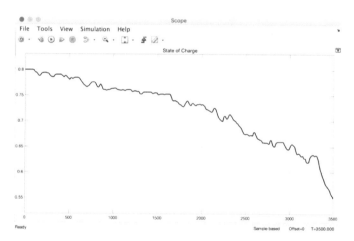

Figure 3.8: State of charge in the battery.

Figure 3.10 shows the equivalent circuit for a single cell in the Supercapacitor block. The circuit is a network of resistors and capacitors that is commonly used to model supercapacitor behavior.

Capacitors $C1$, $C2$, and $C3$ have fixed capacitances. The capacitance of capacitor Cv is variable and depends on the voltage across it. Resistors $R1$, $R2$, and $R3$ have fixed resistances. The voltage across each individual fixed capacitor in the Supercapacitor block is calculated as

$$V_{cn} = \frac{v}{N_{series}} - i_n R_n,\tag{3.6}$$

where:

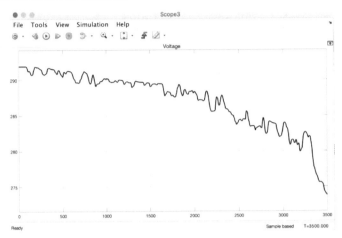

Figure 3.9: Voltage output from the battery.

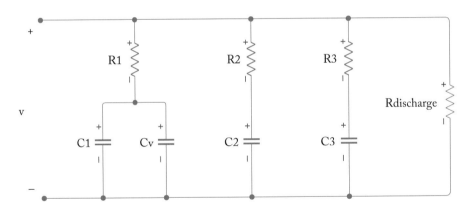

Figure 3.10: Equivalent circuit of the ultracapacitor (supercapacitor).

v is the voltage across the block.

N_{series} is the number of cells in series.

n is the branch number $n = [1, 2, 3]$.

i_n is the current through the nth branch.

R_n is the resistance in the nth branch.

V_{cn} is voltage across the capacitor in the nth branch.

The equation for the current through the first branch of the supercapacitor depends on the voltage across the capacitors in the branch. If the capacitors experience a positive voltage, that is $V_{c1} > 0$.

Then

$$i_1 = (C_1 + K_v V_{c1}) \frac{dV_{c1}}{dt} \tag{3.7}$$

else

$$i_1 = C_1 \frac{dV_{c1}}{dt}, \tag{3.8}$$

where:

V_{c1} is voltage across the capacitors in the first branch.

C_1 is the capacitance of the fixed capacitor in the first branch.

K_v is the voltage-dependent capacitance gain.

i_1 is the current through the first branch.

For the remaining branches, the current is defined as

$$i_n = C_n \frac{dV_{cn}}{dt}, \tag{3.9}$$

where:

n is the branch number. $n = [2, 3]$.

C_n is the capacitance of the nth branch.

The total current through the Supercapacitor block is

$$i_n = N_{parallel} \left(i_1 + i_2 + i_3 + \frac{v}{R_{discharge}} \right), \tag{3.10}$$

where:

$N_{parallel}$ is the number of cells in parallel.

$R_{discharge}$ is the self-discharge resistance of the supercapacitor.

i is the current through the supercapacitor.

We will now put together a fairly simple model to understand how the ultracapacitor model works. Here are the steps to follow.

1. Type ssc_new in the Matlab command window to open a new model file.

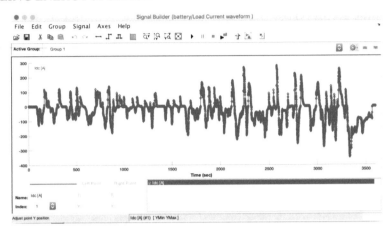

Figure 3.11: Load current waveform.

2. Select and place the Load Current waveform that was used in the previous example. The current profile that is modeled by this block is shown in Figure 3.11. Also add a constant block in the model and use an add block to add the step function to the load current block. Set the constant value to 25. Add a Goto block and name it Iin. Connect the total current to this Goto block.

3. Add a Supercapacitor block, a controlled current source, a voltage sensor, an electrical reference, a S-PS converter, and a PS-S converter.

4. Take the Iin and connect it to the S-PS converter and then into the current source block. Connect the output of the current source block to the supercapacitor. And connect the voltage sensor across the supercapacitor. Connect the reference block to complete the circuit as shown in Figure 3.12. Set the unit on the S-PS block to mA.

5. Connect a PS-S block to the output of the voltage senor and store the output in the Goto block named V.

6. Add a Scope and connect two From blocks to the scope. Name the From blocks Iin and V, respectively. The entire model looks as shown in Figure 3.12.

7. Run the simulation for 3500 s.

8. Figure 3.16 shows the output results.

This current supplied is a replication of current profile in a stop and go traffic, it switches from positive to negative and back over very short periods of time. The ultracapacitor charges and discharges during that period and the voltage measured across it accordingly fluctuates as is seen in Figure 3.16.

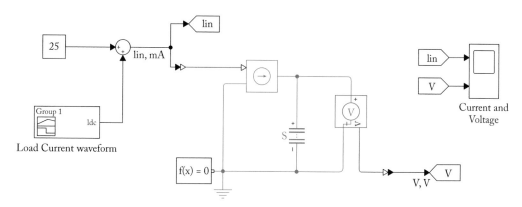

Figure 3.12: The ultracapacitor model.

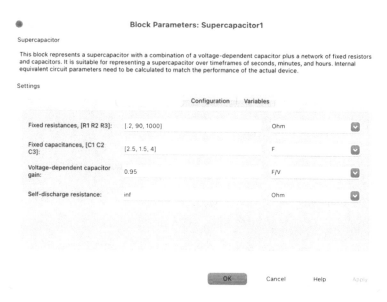

Figure 3.13: Supercapacitor's cell characteristics settings.

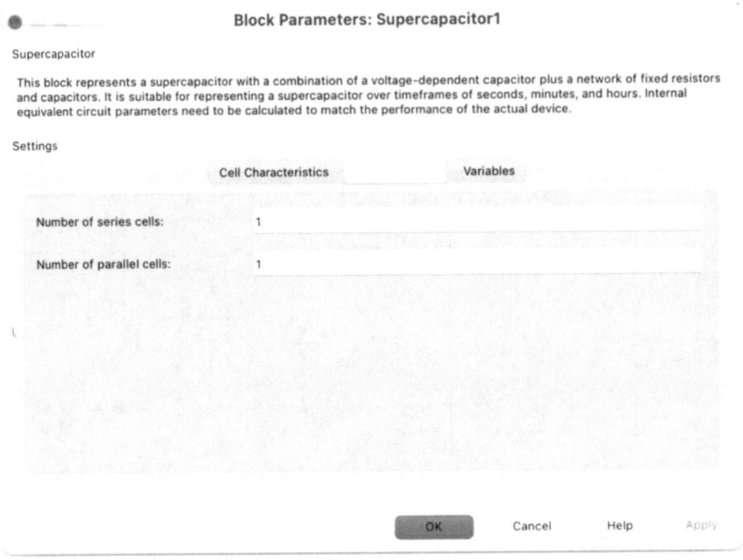

Block Parameters: Supercapacitor1

Supercapacitor

This block represents a supercapacitor with a combination of a voltage-dependent capacitor plus a network of fixed resistors and capacitors. It is suitable for representing a supercapacitor over timeframes of seconds, minutes, and hours. Internal equivalent circuit parameters need to be calculated to match the performance of the actual device.

Settings

Cell Characteristics	Variables

Number of series cells:	1
Number of parallel cells:	1

OK   Cancel   Help   Apply

Figure 3.14: Supercapacitor's configuration setting.

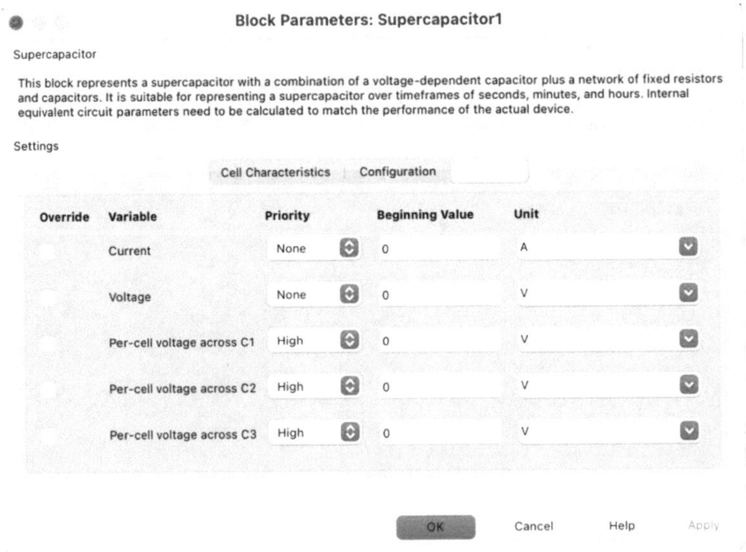

Block Parameters: Supercapacitor1

Supercapacitor

This block represents a supercapacitor with a combination of a voltage-dependent capacitor plus a network of fixed resistors and capacitors. It is suitable for representing a supercapacitor over timeframes of seconds, minutes, and hours. Internal equivalent circuit parameters need to be calculated to match the performance of the actual device.

Settings

Cell Characteristics	Configuration

Override	Variable	Priority	Beginning Value	Unit
	Current	None	0	A
	Voltage	None	0	V
	Per-cell voltage across C1	High	0	V
	Per-cell voltage across C2	High	0	V
	Per-cell voltage across C3	High	0	V

OK   Cancel   Help   Apply

Figure 3.15: Supercapacitor's variables setting.

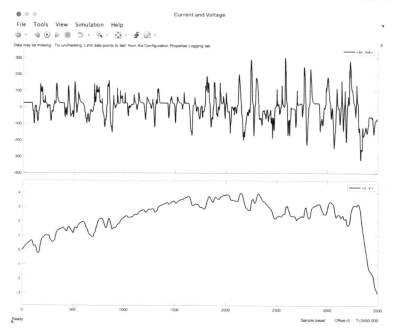

Figure 3.16: Current supplied through the ultracapacitor and voltage across the ultracapacitor.

3.3   SUMMARY

In this chapter we discussed the models for the two main types of energy storage devices that are used in EV and HEVs: the battery and the ultracapacitor. In later chapters, we are going to use these models in conjunction with models of other parts of the vehicle drive train to establish more details of the EV and HEV drivetrain.

CHAPTER 4

Modeling DC Motors and Their Control

4.1 INTRODUCTION

The Electric Drive is a key sub-system of any HEV or EV system. The term Electric Drive is usually used for the traction motors and the associated power electronics and controls that ensures a desired motor performance. In this chapter we will start by considering some of the DC motors that could be good candidates for traction motor and explore their behavior and characteristics. In the next two chapters this discussion will continue when we consider power electronics in Chapter 5 and continue to consider some relevant AC motors in Chapter 6.

There are several types of DC motors. The simplest one is the Permanent Magnet DC motor where the magnetic field is generated using a permanent magnet. There are a many others such as the separately excited DC motor, the series DC motor, the shunt motor, and a series-parallel combination motor. In all of these later ones, the magnetic field is generated by an electromagnet or a coil through which an electric current is supplied. In a separately excited DC motor, the coil that generates the magnetic field (stator) and the armature coil (rotor) are two different ones (hence the name "separately excited"). In the series, shunt and the combination motors, the two coils are electrically connected and the names indicate the nature of the connection as in the two coils are connected in series, in parallel, and in a series–parallel combination mode. Due to this electrical coupling, the field and armature currents are not completely independent of each other, unlike the separately excited DC motor. It so happens that of all these types of DC motors only two, the Permanent Magnet DC and the Separately Excited DC, are good candidates as traction motors and our discussion in this chapter will be limited to those two only. Although they are not used in today's HEV or EV automobiles due to some significant drawbacks, their behavior and characteristics have key features that are necessary for traction motors and can be replicated in AC motors that have ultimately become the motors of choice for traction applications.

4.2 PERMANENT MAGNET DC MOTOR

Figures 4.1 and 4.2 show schematics for a permanent magnet DC motor. This is the simplest DC motor where the magnetic field is created by a set of permanent magnets that act as the stator and the armature windings are the rotor. Permanent magnet provides a constant magnetic field

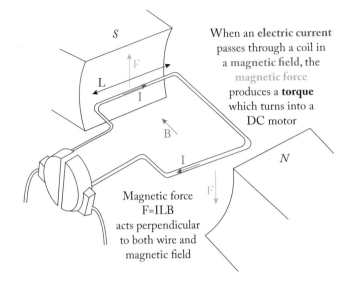

Figure 4.1: A schematic to show the interaction of a current-carrying conductor and a magnetic field.

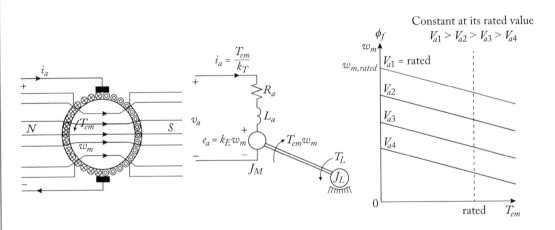

Figure 4.2: Schematic of a motor cross-section, the motor circuit, and typical torque vs. speed curves for a PMDC motors.

or flux density. For an armature conductor of length L carrying a current I, the force resulting from a magnetic flux density B at right angles to the conductor is BIL (Figure 4.1). With N conductors the force is $F = NBIL$. The forces result in a torque about the coil axis of Fc, if c is the diameter of the armature (or the distance between a pair of opposite forces, as shown in Figure 4.1). So, the torque may be written as $T = (NBLc)I$. Torque is thus written as $T = K_T I$;

I = armature current, and K_T is a constant based on motor construction. The armature coil is rotating in a magnetic field, electromagnetic induction will occur and a back emf will be induced. The back emf E is related to the rate at which the flux linked by the coil changes. For a constant magnetic field, this is proportional to the angular velocity of rotation. Hence, back emf is related to flux and angular speed, $E = K_E \omega$; where ω = motor speed in rad/s.

K_T and K_E depend on motor construction and they are of the same magnitude (but of different units). Armature current, at steady state, is (because the armature inductance behaves like a connecting wire at steady state) $I = (V - E)/R$. R is the armature resistance and E is back emf. The Torque therefore is $T = T = K_T I = K_T (V - E)/R = K_T (V - K_E \omega)/R$. At start-up, back emf is minimum, therefore, I is maximum and Torque is maximum. The faster the motor runs the smaller the current and hence the torque. The motor circuit is shown in Figure 4.2. The current in the circuit is $I = (V - E)/R$ at steady state.

When one considers the unsteady state situation the rate of change of current at the initial state needs to be considered (due to the presence of the inductor). This results in a system of two coupled first-order equations (one each on the Electrical and Mechanical side). All the motor equations together are:

$$
\begin{aligned}
\frac{dI}{dt} &= \frac{1}{L}(V - E - IR) \\
\frac{d\omega}{dt} &= \frac{1}{J}(T - T_L - B\omega) \\
E &= K_E \omega \\
T &= K_T I.
\end{aligned}
\tag{4.1}
$$

The first two equations are the governing differential equations and the next two are the relationships due to the electro-magnetic interaction. The system parameters are: L (*armature inductance*), J (*rotor inertia*), R (*armature resistance*), B (*rotational damping*), $K_T = K_E$ (*torque or voltage constant*), and T_L (*load torque*).

4.3   EXAMPLES

Following are a handful of tips to add new elements into a model.

Tips for adding modal elements

1. Use Quick Insert to add the blocks. Click in the diagram and type the name of the block. A list of blocks will appear and you can select the block you want from the list. Alternatively, the Open Simscape Library block can be used to look though the library of all blocks and pick the appropriate one.

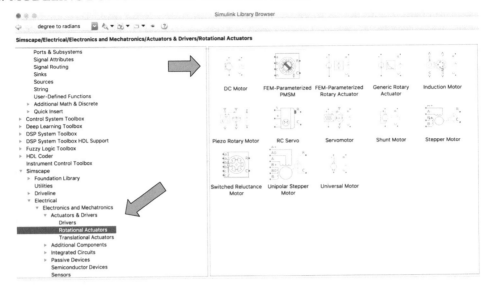

Figure 4.3: Locating the DC motor block in the electrical library.

2. After the block is entered, a prompt will appear for you to enter the parameter. Enter the variable names as shown below.

3. To rotate a block or flip blocks, right-click on the block and select Flip Block or Rotate block from the Rotate and Flip menu.

4. To show the parameter below the block name, see Set Block Annotation Properties in the Simulink documentation.

4.3.1   EXAMPLE 4.1: PERMANENT MAGNET DC MOTOR MODEL

Simscape offers a DC Motor block within its libraries that incorporates most of the electrical as well as mechanical elements of the motor within this one block. In this example, the DC Motor block is used to model a permanent magnet DC machine.

Beyond the foundation library, Simscape has a number of specialized libraries that contain specialized blocks that may be used to model many advanced engineering systems. Figure 4.3 shows us accessing one such library to retrieve the DC motor block. Within Simscape there is a library called Electrical and within Electrical there is a sub directory for Electronics and Mechatronics. The DC-Motor model along with some other motor models are available in the Rotational Actuators subdirectory of the Actuators and Drivers directory.

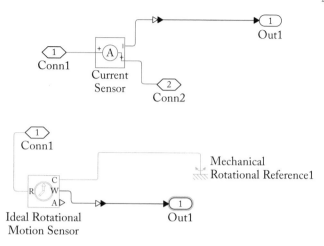

Figure 4.4: Two sensor subsystems.

Steps:

1. Type **ssc_new** in the Matlab command window to open a new model file.

2. Select and place a Current Sensor, an Ideal Rotational Motion Sensor.

3. Using each sensor create a sensing subsystem in the following way: Select an Outblock from Simulink library and a PS_S converter. Connect the output of the sensor through the PS_S block to the Out1 block. Select all these items, right-click, and choose to create a subsystem from the menu. Once the subsystem is created, the software automatically adds a Conn1 and Conn2, two connectors for the subsystem to the C and R terminals of the sensor. These connectors will be used to connect this subsystem at appropriate locations in the model. The current sensor is going to measure a Through variable. The Rotational motion sensor will measure an Across variable so use a Mechanical rotational reference block to connect to the C terminal for the rotational motion sensor. The R terminal will be connected to the motor output. Figure 4.4 shows these two sensor subsystems.

4. Name these two subsystems, current measurement and speed measurement, respectively.

5. Since we will be tracking and plotting a number of variables in the model we would like to minimize the clutter in the model so that it looks clean. The From and Goto blocks available in the Simulink library are useful blocks to read and write data. These can be used to store data at one location in the model and retrieve the same data at another location in the model. Add two Goto blocks and two From blocks in the model. The output of each sensor will be stored in the Goto blocks and they will be retrieved from the From blocks for plotting. For bookkeeping purposes, the Goto and From blocks that correspond to

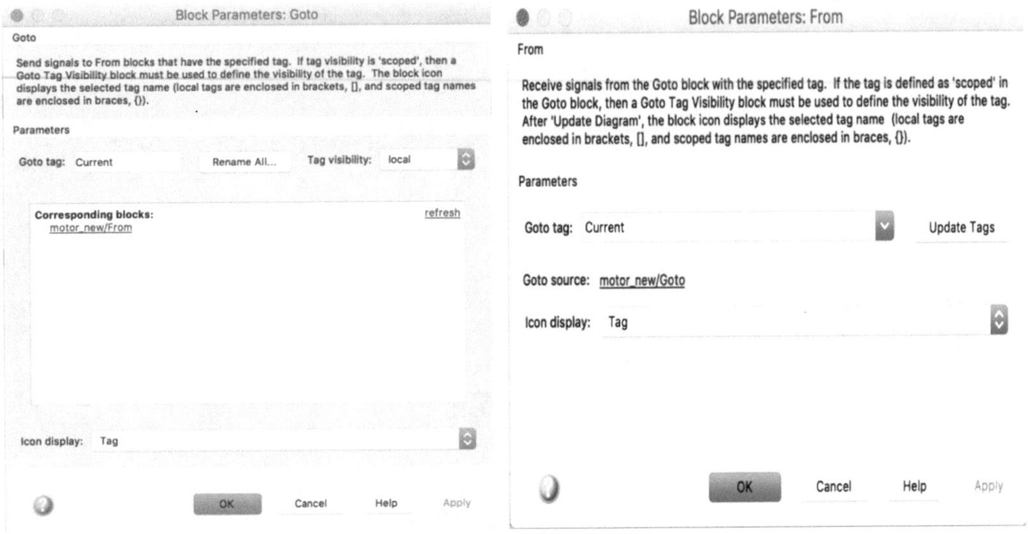

Figure 4.5: Goto and From block setups.

each other should have the same variable name. Double-click on the first Goto block and set the tag to Current and connect it to the current sensor subsystem output. Double-click on the first From block and set the tag to Current and connect it to a scope. Name the scope current. Figure 4.5 shows the setup menus for these blocks. Now these Goto-From block pairs will track the current in the circuit and plot it on a scope. Do the same for the other two pairs and name them Speed. Connect them to the appropriate sensor subsystem outputs.

6. Now we will add all the main elements in the model. Select and place the Controlled Voltage Source, Electrical Reference and Mechanical Rotational Reference, and a Step Input block.

7. Add the DC Motor block from the electrical library, as shown in Figure 4.3.

8. Connect the elements to create the model shown in Figure 4.6. Notice that the electrical side elements of the model has a different color from the mechanical side.

9. Open the DC Motor setup menu and set necessary motor parameters both for electrical and mechanical components. See Figures 4.7 and and 4.8. The parameters are set to match the values used in the previous example (Motor Inductance = 1e-6 H, Motor Resistance = 1 Ohm, Rotational Intertia = 0.01 kgm^2, Damping coefficient = 0.001 N/rad/s, Motor constant (torque and back emf) = 0.1 V/rad/s = 0.1 Nm/A). There is no additional load torque.

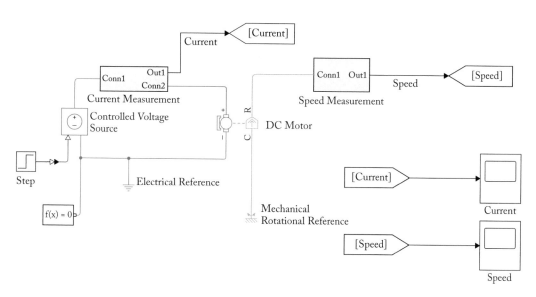

Figure 4.6: Motor model with a DC motor block.

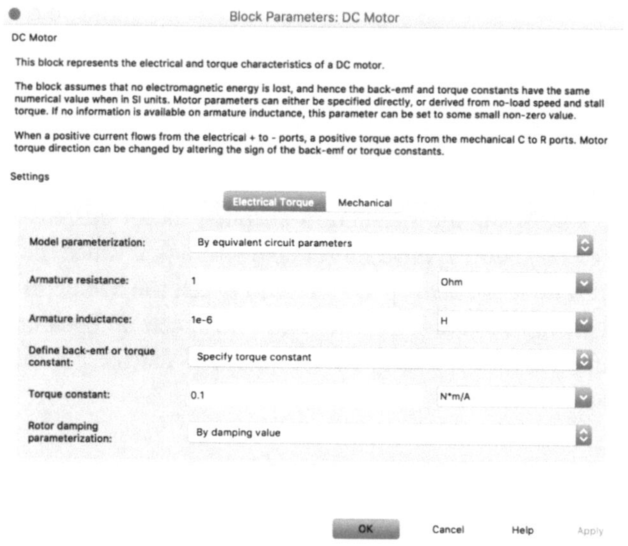

Figure 4.7: Setting the electrical parameters.

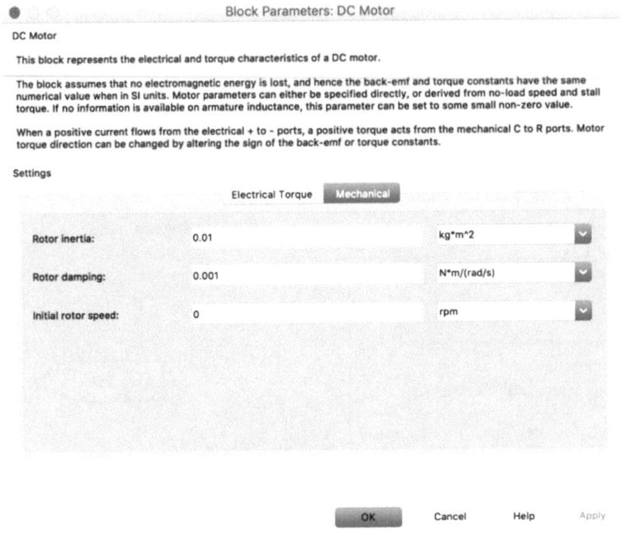

Figure 4.8: Setting the mechanical parameters.

10. The Step input setting of the step input block should be set at 10 after a time of 1 s. This applies a voltage of 10 V, 1 s after the simulation start time.

11. Use the ODE15s stiff ODE solver by choosing it in the Model Configuration option in the Simulation menu.

12. Run the simulation for 10 s.

Figures 4.9 and 4.10 show the simulation outputs, motor current and motor speed, respectively. The results clearly show typical motor characteristics. Once the 10 V input is applied the motor starts to speed up from zero speed. At zero speed the back emf is zero and the hence the current drawn is the highest. As the motor starts to speed up back emf increases and consequently the motor current drops. With the motor speed reaching a steady value the current settles to a steady value as well.

4.3.2   EXAMPLE 4.2: CONTROL OF DC MOTOR FOR SPEED

In this example we will modify the motor model from the first example to add a control loop for speed control tasks. We will use a tunable PID controller provided by the Simulink library.

Steps:

1. Open the motor model from Example 4.3.1 (Figure 4.6).

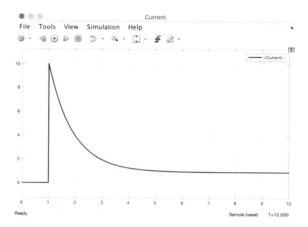

Figure 4.9: Current in the motor.

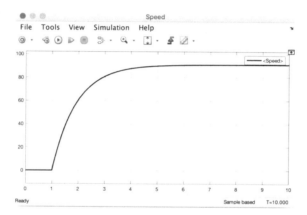

Figure 4.10: Motor speed.

2. Add a In1 block and an Out1 block to the model from the Simscape library. Remove the step input in the model and connect the In1 block to a controlled voltage source. Connect the Out1 block to the motor speed output that was connected to the scope.

3. Select everything on the screen and right-click to see the menu and choose to build a subsystem. This creates the motor subsystem with one input and one output. Name this subsystem Motor_Model.

4. Now add a Step Input, an addition-subtraction block, a scope, and a PID controller block from the Simulink Library.

5. Connect the elements to create the model shown in Figure 4.11.

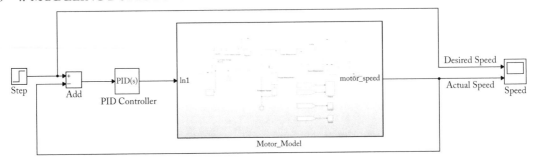

Figure 4.11: Motor speed control model.

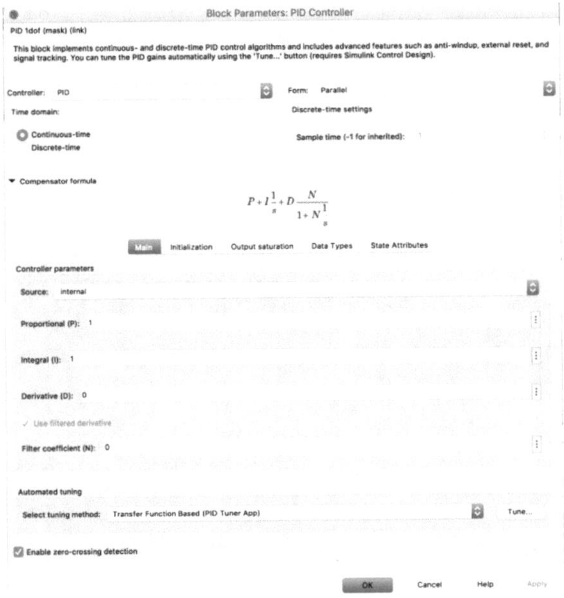

Figure 4.12: PID controller block parameter menu.

6. The Step input provides the desired speed and that subtracts the actual speed to create the error signal. The error signal is passed through the PID controllers to develop the control signal. The control signal that the controller generates is fed into the Motor Model as the voltage input to the model. The actual speed output from the model and the desired speed are plotted together for visualization and judging the effectiveness of the model.

7. Double-click on the PID controller to open the block parameter menu. By default, it is a PID controller. We can choose to make it a P, PI, or PD controller by picking the appropriate setting at the top-left corner. By default, the unit values are assigned to P and I controller parameters and zero for D parameter. Figure 4.12 shows the block parameter menu. We will use these default values to explore the behavior of the controller.

8. Ensure that the ODE15s is the solver chosen in the Configure parameters menu under simulation and run the simulation.

9. Figure 4.13 shows the desired speed, a step change from 0–10 at 1 s, and the actual speed. With this initial choice of controller parameters the controller seems to be working pretty well.

10. To improve the performance of the system the controller is changed to a PI controller and the P value is changed to 20 and the I value is set to 1. And the system is re-analyzed. The speed comparison is shown in Figure 4.14.

11. If the armature current is tracked (as shown in Figure 4.15) it shows a sharp spike at the 1 s mark where the speed change happens. The sharp spike in this case goes as high as 200 A. This could be unacceptable in a given situation. Such spikes, if occurring repeatedly or for any significant length of time, could lead to motor damage and/or burnout.

12. A standard approach to reduce these current spikes is to use multi-level controllers, also called cascade control. For a two-level cascade controller two controller blocks are used, the outer one is for speed control and the inner is for current or torque control. For this control strategy to work two signals, the speed and the current, are used in the feedback loop. The DC motor model with cascade controllers is shown in Figure 4.16. For the current controller the P value is picked as 0.3 and the I value is picked as 0.125.

13. Figure 4.17 shows a comparison of desired and actual speeds and Figure 4.18 shows the current profile. While the speed results remain largely similar to the speed profile with the single controller, the current peak has dropped down significantly from the 200 A value to about 45 A. The use of cascade controllers remains an effective way to control current spikes in motors. With more fine tuning of the controller parameters (P and I values) even better results can be obtained. Controller tuning is an important aspect of controller design but no formal discussion of that aspect is attempted in this book.

4.4   PMDC MOTORS AS A TRACTION MOTOR

Design, construction, and control of Permanent Magnet DC motors are simple and easy to understand and implement which is why these motors were considered for traction or drive applications, but were quickly rejected due to some very critical drawbacks. The primary drawback is that only a single input can be used to control these motors, i.e., since the magnetic field is fixed, the only variable that can be changed to change the motor response is the armature voltage/current. So, if the motor output is to be changed, the user can only increase or decrease the armature voltage. To increase the speed of the motor one has to increase the applied voltage, which means the current drawn will keep increasing. This cannot happen in an unlimited fashion. With the increase in current the heat generated in the armature will increase proportional

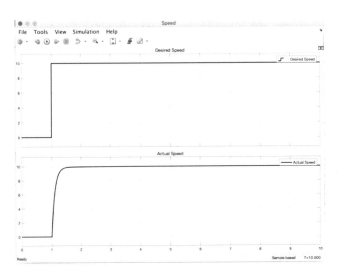

Figure 4.13: Comparison of desired and actual speed with initial settings.

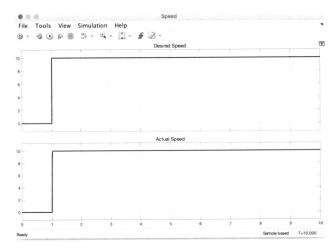

Figure 4.14: Desired and actual speed comparison with improved P value.

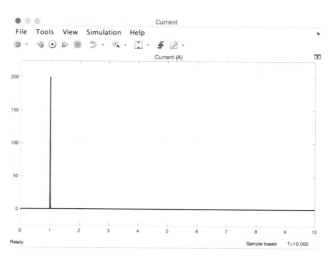

Figure 4.15: Armature current showing a sharp spike.

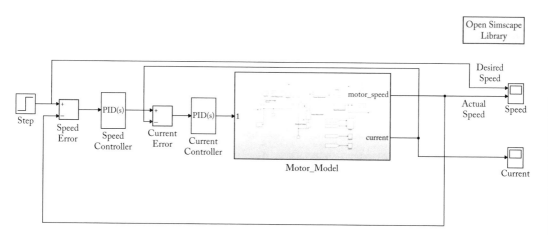

Figure 4.16: Motor model with cascade controller.

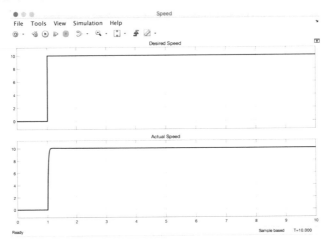

Figure 4.17: Speed comparison with cascade controller.

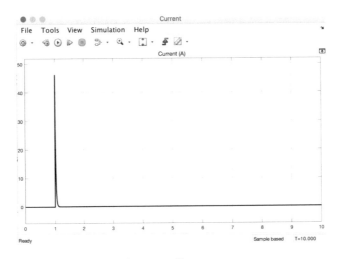

Figure 4.18: Current profile with cascade controller.

to the square of the current. Very quickly a higher and higher amount of heat generation will lead to temperature rise and armature burn-out. In the case of DC motors, since the armature is at the core of the motor and there is a airgap between the armature and the stator, the core is in general not well designed for heat dissipation by either conduction or convection. This is why PMDC motors become unfeasible for large capacity or large size applications. They are better suited for small size applications, such as wiper or window motors or other mechatronic applications. Besides this primary reason of heat dissipation problems, friction between commutator rings and carbon brushes is also a problem due to frictional wear and tear and need for frequent maintenance. Separately excited DC motor is an improvement over PMDC motors as far as the issue of overheating and single-signal control is concerned. We will now consider a model of separately excited DC motors.

4.5   SEPARATELY EXCITED DC MOTORS

In a separately excited DC motor, the magnetic field is created by an electromagnet (field coil) instead of a permanent magnet. The construction and design of the rotor is essentially the same as a PMDC motor including the carbon brushes and commutator rings to switch the direction of motor current. Figure 4.19 is a schematic of the separately excited DC motor. The motor equations remain very similar to the PMDC motors (Equation (4.1)). They are shown here as Equations (4.2)–(4.7). There are two electrical circuit equations (for the field and the armature circuits). The rate of change of currents in each of those circuits are expressed in Equations (4.2) and (4.3). Subscript a is used for the armature circuit and subscript f is used for the field circuit. L_f and L_a are field and armature inductances, respectively, and u_f and u_a are supply voltages for the field and armature circuits, respectively; r_f and r_a are the resistances in the two circuits and i_f and i_a are the currents in the two circuits, respectively. Equation (4.4) is the motor speed equation where ω_r is the angular speed of rotation. Equations (4.5) and (4.6) link the electrical and mechanical sides through the Torque and back emf equations. For the PMDC motor the torque was written as $T = (NBLc)I$. This resulted in $T = K_T I$; $I =$ armature current, K_T was a constant based on motor construction because all the other items in the equation are constants. In the case of the separately excited DC motor, the torque expression is essentially the same with one exception, B the magnetic flux density is not a constant but is proportional to the field current. So the torque equation can be written as $T = (NBLc)I = [(NLc)ki_f]i_a = (Ki_f)i_a$, where i_f is the field current and i_a is the armature current and K is a constant that captures all other parameters in the equation, and Ki_f is equivalent to the torque constant in the PMDC motor. Extending this concept, the back emf $e = (Ki_f)\omega_r$. Equation (4.7) is the steady-state torqe vs. speed equation of the separately excited DC motor. From this equation it is clear that there are two quantities that can be varied to change the motor performance, namely the armature voltage (and therefore the armature current) and the field current (therefore the field voltage). The typical processes are as follows: after the motor is started, the armature voltage/current is increased to increase the motor speed and torque. As the motor speeds up, the torque output increases (as does the armature current) until the motor reaches its rated base speed and rated power output (the envelope is shown as the horizontal line between the origin and location 1 on the graph in Figure 4.20). Once the rated base speed is reached the controller is switched to field control. The field control region is also called the constant power region. Here, the field voltage/field current is reduced and this weakens the magnetic field and as a result, speeds up the motor (the field current is the denominator in Equation (4.7)). During this portion of the operation the armature voltage is held constant. In the following model this entire control process for the separately excited DC motor is replicated.

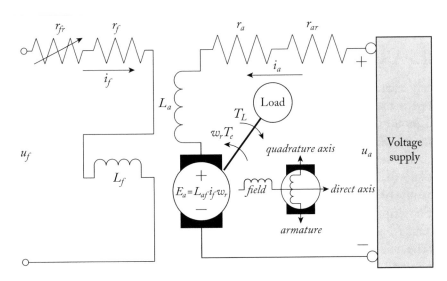

Figure 4.19: Separately excited DC motor schematic/circuit.

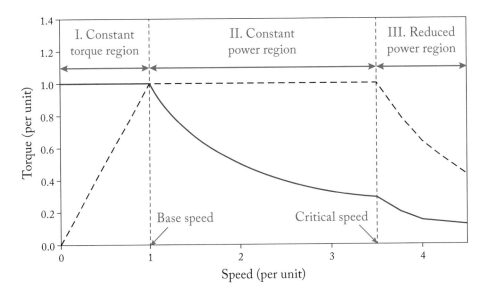

Figure 4.20: Separately excited DC motor operation envelope.

$$\frac{di_f}{dt} = -\frac{r_f}{L_f}i_f + \frac{1}{L_f}u_f \tag{4.2}$$

$$\frac{di_a}{dt} = -\frac{r_a}{L_a}i_a - \frac{K}{L_a}i_f\omega_r + \frac{1}{L_a}u_a \tag{4.3}$$

$$\frac{d\omega_r}{dt} = \frac{1}{J}\left(T_{em} - T_{viscous} - T_L\right) = \frac{1}{J}\left(Ki_ai_f - B\omega_r - T_L\right) \tag{4.4}$$

$$T_{em} = (K_a\phi)\,i_a = Ki_fi_a \tag{4.5}$$

$$e_a = (K_a\phi)\,\omega_r = Ki_f\omega_r \tag{4.6}$$

$$\omega_r = \frac{u_a - r_ai_a}{Ki_f} = \frac{u_a}{ki_f} - \frac{r_a}{(Ki_f)^2}T_e. \tag{4.7}$$

4.5.1   EXAMPLE 4.3: SEPARATELY EXCITED DC MOTOR WITH ARMATURE AND FIELD CONTROL

Simscape offers many motor models in its library of models. Beyond the foundation library, Simscape has a number of specialized libraries that contain specialized blocks that may be used to model many advanced engineering systems. Figure 4.21 shows us accessing one such library to retrieve a model that can be used as a separately excited DC motor block. Within Simscape there is a library called Electrical and within Electrical there is a sub directory for Specialized Power Systems. The DC-Machine model along with other motor models are available in the Machines subdirectory of the Fundamentals directory within Specialized Power Systems.

Steps:

1. Open a new Simscape model and add the DC Machine icon. The separately excited model needs two sources, one for the armature and one for the field. Add two Controlled Voltage sources. These can be configured for AC or DC sources and can supply variable values with appropriate input. Connect these two to the field and the armature inputs and a ground link, as shown in Figure 4.22.

2. Add three input blocks and connect them to the load torque input port of the DC machine, and the two supplies, respectively. Rename them Load Torque, Field Voltage, and Armature Voltage, respectively.

3. The output from the DC Machine can be tracked from the "m" port. There are four outputs that can be monitored at this port: motor speed, armature current, field current, and output torque. Add a Demux block and set the number of outputs to 4. Connect the output from m to the Demux block and connect four outport blocks and name them appropriately to identify the signal they are carrying. The output speed from the m port is in rad/s so add a

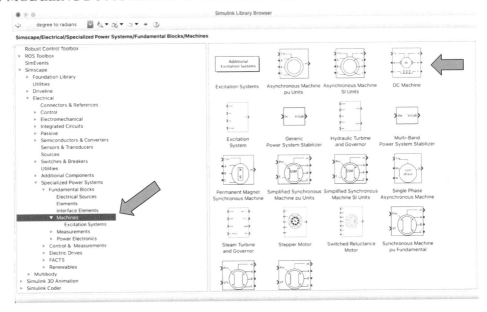

Figure 4.21: Locating the DC machine block in the electrical library.

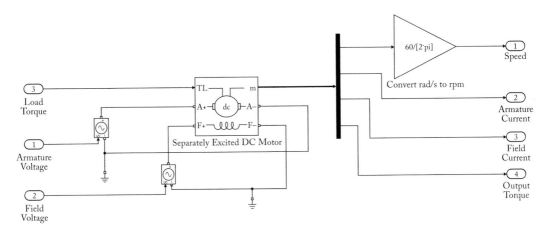

Figure 4.22: The motor subsystem.

gain block to convert the speed from rad/s to rpm (multiply by 60/2pi). The Motor model should look like Figure 4.22.

4. Open DC Machine settings by double-clicking and change the setting, as shown in Figure 4.23. This block allows the use of preset DC machines. We pick a motor with specification of 5 hp, 240 V, 1750 rpm, and field voltage of 300 V. Here, 240 V is the rating of

Figure 4.23: DC machine settings.

the Armature and 300 V is the rating of the field voltage. 1750 rpm is the speed limit at the maximum Armature voltage and at the maximum power output. To achieve even higher speed flux weakening will have to be used. We will be using load torque as the mechanical input so that can be set here as well. Choose everything on screen in the motor model as in Figure 4.22 and create a subsystem. Name it Motor System.

5. Add a Signal Builder block and create a signal that starts at 0 and increases linearly from zero to 2000 in 15 s. It stays constant at 2000 from 15–20 s and steadily reduces to zero from 20–35 s. Name this Signal Builder "Speed-Limit." Connect a Goto block to the signal builder and name it Desired_speed.

6. Add a Goto block and name it speed. Connect it to the speed output port of the Motor Model. Also, connect a scope to that same output. Connect three more scopes to the Armature_Currrent, Field_Current, and the Output_Torque ports.

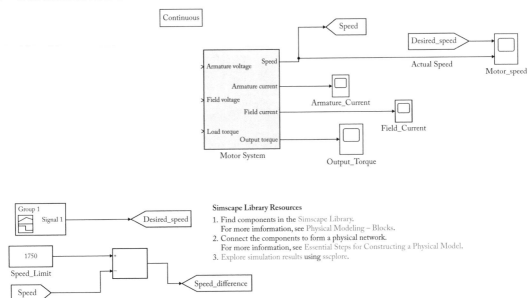

Figure 4.24: Separately excited DC motor model and inputs and outputs.

7. Add a Add block and add a constant block and a From block to be subtracted from the constant. The output of the plus-minus block is connected to a Goto block and name it speed_difference. Name the From block Speed and the constant block Speed_limit. Set the Speed_limit constant value to 1750. Set all the Goto and From blocks used so far to be global variables. Figure 4.24 shows the model as it is setup so far. Now we will add additional elements to implement the control algorithm that will switch between armature control and field control.

8. Add a new subsystem and name it Control_System. Open this subsystem and add two Add-subtract blocks, two Switch blocks, two PID controller blocks, five From blocks, and two Constant blocks. These will be used to develop the two control strategies. We know that until the speed of the motor reaches 1750 rpm (the rated speed) the motor will be armature controlled, i.e., the armature current will be increased to speed up the motor. Once the motor reaches the rated speed and additional speed increase is necessary it will be achieved by Flux weakening. At that point the armature voltage will be held at the steady value of the rated voltage and field voltage will be adjusted to achieve the higher speed through flux weakening.

9. A switch block needs three inputs. The middle one is the condition. When this middle input is greater than 0 the first signal is passed through and when the middle input is less than zero the third signal is passed through. The output of the first switch is fed to

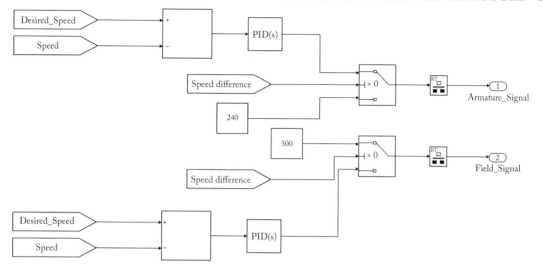

Figure 4.25: Controller sub-model to implement armature and field controls.

the armature as armature voltage and the output of the second switch is fed to the field coil as the field voltage. The logic is as follows: the Speed_difference parameter is the difference between 1750 (rated speed) and the actual motor speed. This is the condition for middle input for both switches. When this is positive, the motor is still in the armature control mode and the upper switch is passing through a PID-controlled armature input and the lower switch passes through 300 V to the field coil (the rated voltage for the field to generate the maximum field flux). As soon as the speed starts to go past 1750 rpm, the upper switch switches the armature supply to 240 V, the armature rated voltage, and the lower switch passes a PID signal to the field coil. In both cases the PID signal is based on the speed error (actual speed-desired speed). The outputs for both switches are passed through a rate transition block that helps smooth out the rate of data used by two sides of a model. Figure 4.25 shows the entire control subsystem.

10. For the PID used in the armature circuit the PID values chosen are: $P = 100$, $I = 50$, $D = 0.001$. For the PID used in the field circuit the PID values chosen are: $P = 200$, $I = 40$, $D = 0.001$. No effort was made in this exercise to optimize or fine tune these values. Figure 4.26 shows the entire assembled model. Set the motor torque load input to 20 Nm.

11. Run the simulation for 35 s.

The key output results are shown in Figures 4.27, 4.28, 4.29, and 4.30. Figure 4.27 shows a comparison of the desired speed and the actual speed of the motor. The speed exceeds 1750 rpm around 13.25 s. This is when the field current drops, as shown in Figure 4.28. This part of the

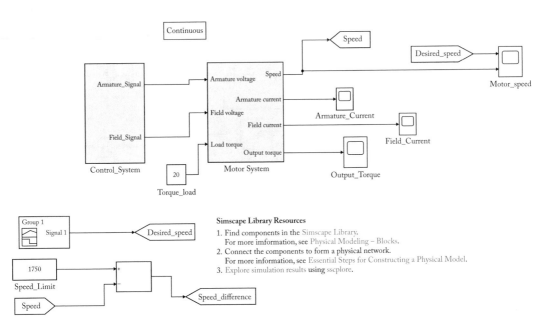

Figure 4.26: Entire model for separately excited DC motor with controller.

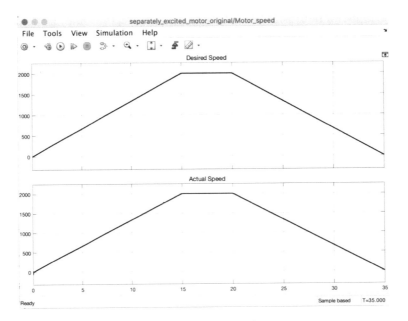

Figure 4.27: Desired speed and actual speed profile for the motor.

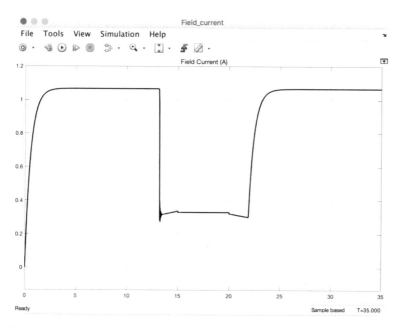

Figure 4.28: Field current for the motor.

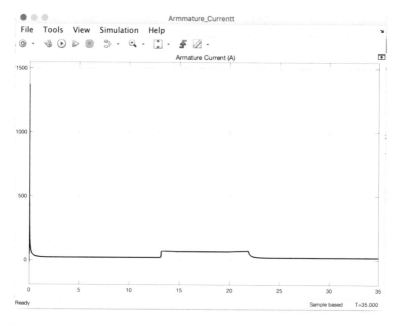

Figure 4.29: Armature current for the motor.

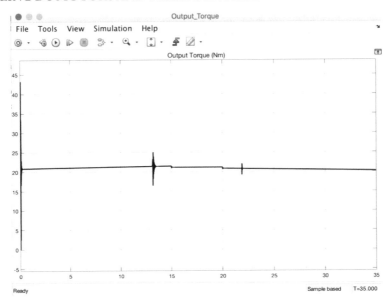

Figure 4.30: Motor output torque.

motor control is implemented through flux weakening. Since the field flux is directly proportional to the field current the dropping of the field current is associated with field weakening and therefore the speeding of the motor. The speed of the motor remains above 1750 until about approximately 22.5 s. When it reaches below 1750 rpm and the field current rises again as the controller switches to armature control. Figure 4.27 shows the armature current, and Figure 4.28 shows the output torque. The output torque is slightly above 20 Nm for the entire time period, and 20 Nm is the load torque. The rest of the torque generated is to overcome frictional losses. This particular motor model has a two-part friction loss, a constant value coulomb friction of 0.5131 Nm, and a viscous friction factor of 0.002953. The viscous friction is proportional to the speed of rotation. The Torque output plot reflects this.

4.6   DC MOTORS AS TRACTION MOTORS: A SUMMARY DISCUSSION

In this chapter we discussed models of two DC motors, the PMDC and the separately excited DC motor. The PMDC motors are simple in their design and construction and could have been a possible candidate for traction applications. There are several reasons for rejecting this motor though. The heat generated in the armature is hard to dissipate for anything more than small motors used for lower power applications. This is because the only way to control these motors is by altering the armature current and higher armature current will eventually generate higher heat. In addition, the use of carbon brushes and commutator rings to switch the current direction leads to the possibility of frequent maintenance needs which is not quite acceptable for vehicular applications. A separately excited DC motor addresses part of the problem with the PMDC motor. There are two ways to control the separately excited DC motor, by changing the armature current and by changing the field current. This way, the armature overheating can be controlled. The problem of brushes and commutator rings (and associated maintenance needs) remains the same in the separately excited DC motors as it was in the case of the PMDC. In the end neither of these are used in the traction applications for HEVs and EVs. A permanent magnet synchronous machine (PMSM), an AC machine, is the most commonly used traction machine today for HEV and EV applications. In this machine the rotor inside the motor is a permanent magnet while the stator is an electromagnet that has three coils and is excited by a 3-phase AC source. The AC source is suitably controlled to replicate the two-pronged control mentioned here, current control as well as the field control. The operational envelope that was introduced in this chapter for the separately excited DC motor (as shown in Figure 4.20) is the same envelope that is used for the PMSM traction machines. The source of energy (the battery) supplies and receives DC power while the traction machine of choice runs on 3-phase AC. This raises the need for a robust power electronics and control module that will constantly transfer DC power to 3-phase AC and vice versa. In Chapter 5, we explore the modeling of some of the key power electronic devices that are used in the various tasks needed for HEV and EV operation. And in Chapter 6 we will discuss modeling the PMSM AC machines.

4.7   SUMMARY

In this chapter we went through the modeling of some DC motors that have characteristics that are favorable for them to be considered as a traction machine for HEV and EV applications. The goal was to establish the modeling process for these machines and highlight their typical characteristics. In actual traction applications AC motors are used and they are controlled in a precise manner so that optimum performance can be ensured for all conditions. In the next two chapters we will consider the modeling of power electronics and AC motors both of which form key aspects of motor drives for HEV and EV applications.

CHAPTER 5

Power Electronics and Hardware Controls

5.1   INTRODUCTION

In Chapter 4, we discussed some of the DC motors which were possible candidates for traction motor. However, none of them can be eventually used due to significant drawbacks in those machines. 3-phase AC motors are used in HEV and EV applications as traction motors. Obviously, AC motors need AC power to drive them. However, the main power source in vehicles is the battery, a DC source. So, there is a need for transforming DC power to AC when driving the vehicle. During battery recharge either by plugging in or through energy recovery from braking the opposite has to happen, AC power has to be transformed to DC. Along with this, there is need to boost and reduce voltage levels that the battery and other devices provide for a variety of scenarios during a vehicle's drive cycle. All of these transferring of power from one type to another as well as vehicle drive control is performed by power electronics.

Power Electronics is a key technical area that plays a very important role in the operation of EVs and HEVs. Some of the core tasks of power electronics involves boosting or reducing voltages, both DC and AC, i.e., raising or lowering voltage levels to values necessary for various parts of the electric drive, converting DC input to 3-phase AC, or to convert AC to DC.

5.2   DIFFERENT TASKS FOR POWER ELECTRONICS

To understand the different tasks that power electronics devices perform in an HEV or EV consider the picture shown in Figure 5.1. This shows a schematic of a series HEV where the power generated by the engine is used to drive the vehicle by energizing an electric motor. Excess generated power is also used to re-charge the battery during part of the operation cycle. The battery is a source and a storage device of DC power whereas the motor and the generator are AC devices. As Figure 5.1 shows, several power electronic devices are used in the HEV system to switch back and forth between AC and DC. The rectifier converts AC to DC, a DC-DC converter steps DC voltages up or down and the inverter converts DC to AC. All the important functionalities of power electronic devices are represented in Figure 5.1.

Based on their functionalities power electronic devices may be divided into the following categories.

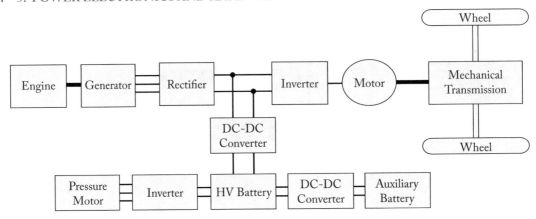

Figure 5.1: **Many power electronics applications in a typical series HEV.**

1. Devices for AC-DC conversion. These are typically referred to as rectifiers and could be single-phase or 3-phase and controlled or uncontrolled.

2. Devices for DC-DC conversion. The different devices are used for voltage Step down (buck convertors), voltage Step up (boost convertors), and versatile devices that can perform both stepping up and stepping down operations (buck-boost convertors, two quadrant choppers or rectifiers, and full bridge converters).

3. Devices for DC-AC conversion. These are typically called inverters. They can be single-phase inverter or multi-phase inverters (typically 3-phase). Also depending on implementation strategies, there are PWM-sine inverters as well as trapezoidal inverters.

4. Devices for AC-AC conversion. These are similar to transformers and we will not discuss much more of these in this text.

5.2.1   MODELING EXAMPLES

The Simscape examples included here are the following:

- AC-DC: Rectifier (half wave), Rectifier (full-wave).

- DC-DC: Boost Convertor, Buck Convertor, DC Motor control for 4-quad operation.

- DC-AC: 3-phase Inverter, 3-phase Inverter with PWM control.

Following are a handful of tips to add new elements into a model.

Tips for adding model elements

1. Use Quick Insert to add the blocks. Click in the diagram and type the name of the block. A list of blocks will appear and you can select the block you want from the list. Alternatively, the Open Simscape Library block can be used to look though the library of all blocks and pick the appropriate one.

2. After the block is entered, a prompt will appear for you to enter the parameter. Enter the variable names as shown below.

3. To rotate a block or flip blocks, right-click on the block and select Flip Block or Rotate block from the Rotate and Flip menu.

4. To show the parameter below the block name, see Set Block Annotation Properties in the Simulink documentation.

5.3   EXAMPLES

5.3.1   EXAMPLE 5.1: AC-DC CONVERSION: RECTIFIER

This example deals with a device circuit that is commonly used in many applications to convert AC input into DC outputs. The circuit is called a rectifier and uses an electronic component, the diode. The diode is essentially an electronic valve that allows flow of current in one direction but stops current from flowing in the opposite direction. This example is a half-wave rectifier; it allows the positive current to pass through and blocks the negative current. It is a good exercise to build this model as a learning tool. However, a more-practical device is a full-wave rectifier which is discussed in the next example.

This example uses an ideal AC transformer and a diode. It first converts 120 volts AC to 12 volts DC. The transformer has a turns ratio of 14, stepping the supply down to 8.6 volts rms, i.e., 8.6*sqrt(2) = 12 volts peak. A resistor is used to represent a typical load.

The following steps are followed to build the model.

1. Type ssc_new in the Matlab command window to open a new model file.

2. Select and place the Resistor, a Diode, Transformer, AC voltage source, and Electrical Reference elements in the workspace.

3. Select and place two Voltage Sensor elements in the workspace.

4. Select and connect two PS-Simulink Converters to the Scope blocks. This block converts the physical signal (PS) to a unit-less Simulink output signal. Connect its input to the outputs of the two voltage sensors. This is the V port, where V stands for "voltage." The V port outputs the voltage as a physical signal which has units. The other ports ("+" and "-") are physical connections to the rest of the circuit.

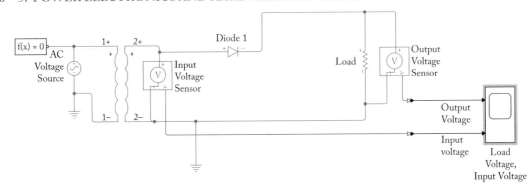

Figure 5.2: Half-wave rectifier model.

5. The Solver Configuration block defines the solver settings for this Simscape physical network. The global default solver settings are used for this model. So leave all boxes unchecked in the solver settings window.

6. Assign parameters for the components: $R = 100$ Ohms, transformer winding ratio $= 14$ and default settings for diode including forward voltage of 0.6 V. Set the input AC voltage frequency to 60 Hz and peak voltage to 120 sqrt(2).

7. Connect the elements as shown in Figure 5.2. Set simulation time to 0.05 s and run simulation.

Figure 5.3 shows a comparison of the input voltage, a full-wave sinusoidal, and the output voltage across the load resistor a half wave/rectified sinusoid. The diode blocks the negative portion of the waveform. The half-wave rectifier is a good tool for conceptualization of the process but full-wave rectifiers are designed using the same concept which are used in many applications where AC to DC conversion is desired. The next example deals with full-wave rectification.

5.3.2   EXAMPLE 5.2: AC-DC CONVERSION: FULL WAVE BRIDGE RECTIFIER

Using[1] an approach similar to the first example, this is a full-wave rectifier model. The full-wave rectifier is a key component in power electronics. The device converts an AC to a DC that is a steady voltage. In this example we discuss how the full-wave rectifier model is built from its circuit. This example uses an ideal AC transformer that was used in the previous example, plus a full-wave bridge rectifier. It first converts 120 volts AC to 12 volts DC. The transformer has a turns ratio of 14, stepping the supply down to 8.6 volts rms, i.e., 8.6*(2) = 12 volts peak. The

---

[1]Note: This example was also used in the text *Modeling and Simulation of Mechatronic Systems using Simscape* by this author and published by Morgan & Claypool.

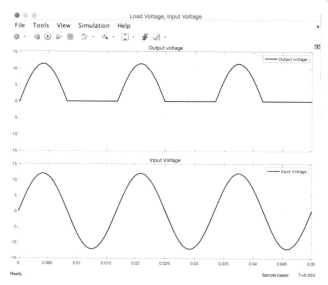

Figure 5.3: Input AC voltage and the rectified half-wave output.

full-wave bridge rectifier plus capacitor combination then converts this to DC. A resistor is used to represent a typical load.

The following steps are followed to build the model.

1. Type ssc_new in the Matlab command window to open a new model file.

2. Select and place the Resistor, Capacitor, four diodes, Transformer, AC voltage source, and Electrical Reference elements in the workspace.

3. Select and place two Voltage Sensor elements in the workspace.

4. Select and connect two PS-Simulink Converters to the Scope blocks. This block converts the physical signal (PS) to a unit-less Simulink output signal. Connect its input to the outputs of the two voltage sensors. This is the V port, where V stands for "voltage." The V port outputs the voltage as a physical signal which has units. The other ports ("+" and "-") are physical connections to the rest of the circuit.

5. The Solver Configuration block defines the solver settings for this Simscape physical network. The global default solver settings are used for this model. So leave all boxes unchecked in the solver settings window.

6. Assign parameters for the components: $R = 100$ Ohms, $C = 470\ \mu$F, transformer winding ratio $= 14$, and default settings for diode including forward voltage of 0.6 V. Set the input AC voltage frequency to 60 Hz and peak voltage to 120 sqrt(2).

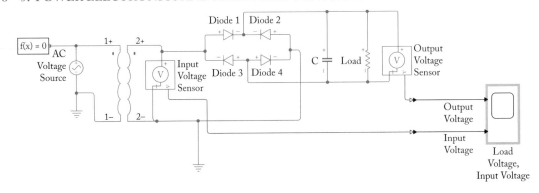

Figure 5.4: Full-wave rectifier.

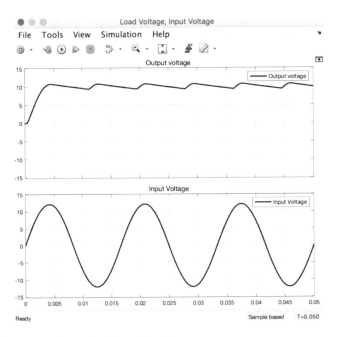

Figure 5.5: Simulation results with $C = 470\ \mu\text{F}$.

7. Connect the elements as shown in Figure 5.4. Set simulation time to 0.05 s and run simulation.

Figure 5.5 shows both the input sinusoidal voltage as well as the output rectified voltage. The DC output voltage has some ripples but is near-steady at 10 V. This model can be used to size the capacitor required for a specified load. For a given size of capacitor, as the load resistance is increased, the ripple in the DC voltage increases. This model is re-run by reducing the capacitor

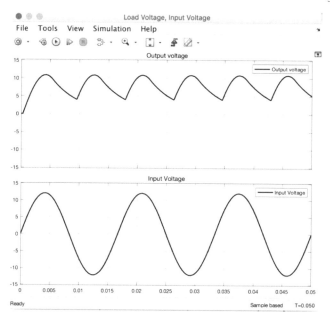

Figure 5.6: Simulation results with $C = 47 \ \mu$F.

size by an order of 10 to, $C = 47 \ \mu$F and the simulation is re-run. The results are shown in Figure 5.6. It is quite clear that the ripples have increased significantly. The model is rerun one more time with $C = 4700 \ \mu$F a 10 times higher value than the original C. Simulation results are shown in Figure 5.7. Ripples have gone down even more than the original but because of the higher C the rise to the steady voltage is slow.

5.3.3   EXAMPLE 5.3: BOOST CONVERTOR: DC-DC CONVERSION

Power electronic circuits are heavily dependent on electronic/semiconductor switches which enable circuit designers to perform fairly complex tasks such as boosting or reducing DC voltages or transforming DC supplies into single phase or 3-phase AC supplies. In EV and HEV applications the source of power is a battery which supplies DC voltage whereas the motor and the drive technology that is used to provide power to the vehicles for vehicle motion runs on 3-phase AC. Power electronic circuits continuously has to switch between AC and DC and therefore these circuits are vital parts of the HEV and EV power trains.

Two electronic switches that are very commonly used in power electronic circuits are MOSFETs (Metal Oxide Semiconductor Field Effect Transistor) and IGBTs (Insulated Gate Bi-polar Transistor). There are many references in open literature about design, construction, and function of these components. We will not get into that discussion here. We will use them to construct models that will demonstrate a particular functionality. We have used the IGBT

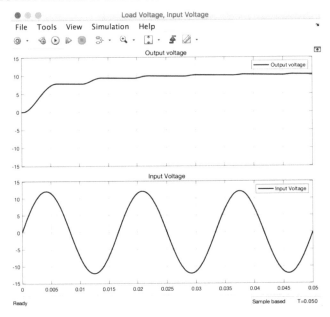

Figure 5.7: Simulation results with $C = 4700 \ \mu F$.

block available within the Electrical subdirectory of Simscape for both examples. In MOSFET and IGBT applications these electronic switches are used by turning them on and off at different frequencies. Based on how they are arranged in a circuit, this allows or stops current along different paths or accumulates energy in storage devices such as an inductor or a capacitor. The specific circuit design then allows the system to provide the desired functionality.

Figure 5.8 shows the model for a Boost convertor. As the name suggests, the goal of this power electronic device is to boost a DC voltage. The supply voltage is from the constant voltage source in the circuit and the output is across the load resistor where a voltage measuring sensor is recording the voltage. The particular functionality of this circuit is achieved by turning the IGBT on and off to alter the current path intermittently. The frequency of IGBT operation is key to achieving the desired boost. All the elements used here are available in the Simscape library either in the fundamental directory or the electrical directory.

Steps:

1. Type **ssc_new** to open a new model.

2. Add an IGBT, a diode, an inductor, two resistors, a DC voltage source, and an electrical reference.

3. Add a pulse generator.

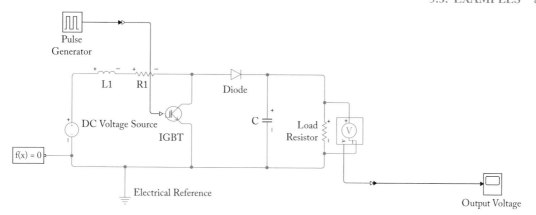

Figure 5.8: Boost convertor circuit model.

4. Set the parameters as: $L1 = 400\text{E-}6$, $R1 = 0.001$ Ohms, DC Voltage $= 100$ V, $C = 25$ E-6 F, Load Resistor $= 60$ Ohms, and a voltage sensor. Connect the elements as shown in Figure 5.8.

5. Figure 5.9 shows the setup menu for the Pulse generator. This shows that the pulse has a period of 100 μs, i.e., the switch on-off frequency is $1/(100\text{E-}6) = 10$ KHz. Also, the switch is on for 50% of the cycle time and off for 50% of the cycle time. This on-off time period is an important parameter in the IGBT behavior and outcome. The boost voltage is given by:

$$V_{boost} = V_{in} \left(\frac{1}{1 - r} \right), \tag{5.1}$$

where r is the fraction of time the switch is on and in this example, it is 0.5. Thus, the output voltage across the Load resistor would be 200 V in this example. Figure 5.10 shows the output voltage to be about 200 V.

5.3.4 EXAMPLE 5.4: DC-DC CONVERSION: BUCK CONVERTER

A buck converter is the opposite of a boost converter, this steps down the output voltage. This circuit also uses a switch such as an IGBT.

Steps:

1. Type **ssc_new** to open a new model.

2. Add an IGBT, a diode, an inductor, two resistors, a DC voltage source, an electrical reference, and a voltage and current sensor.

3. Add a pulse generator.

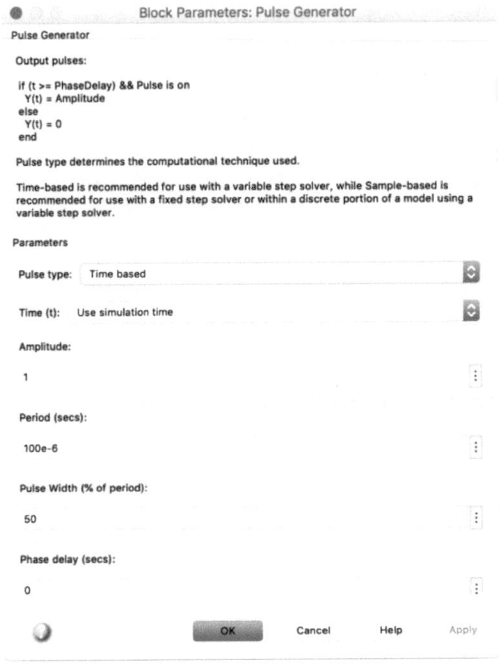

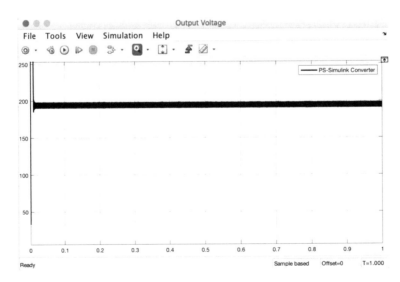

Figure 5.9: Pulse generator setup.

Figure 5.10: Output voltage in a boost convertor.

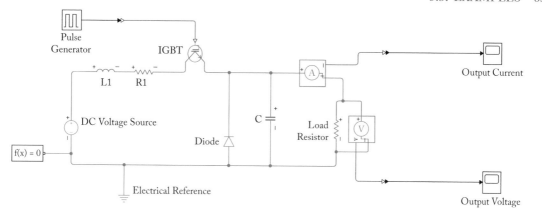

Figure 5.11: Boost convertor circuit model.

4. Set the parameters as: $L1 = 375\text{E-}6$, $R1 = 0.001$ Ohms, DC Voltage $= 100$ V, $C = 125$ E-6 F, Load Resistor $= 50$ Ohms.

5. Connect the elements as shown in Figure 5.11.

6. Figure 5.12 shows the setup menu for the Pulse generator. This shows that the pulse has a period of 100 μs, i.e., the switch on-off frequency is $1/(100\text{E-}6) = 10$ KHz. Also, the switch is on for 50% of the cycle time and off for 50% of the cycle time. This on-off time period is an important parameter in the IGBT behavior and outcome. The boost voltage is given by:

$$V_{boost} = rV_{in}, \qquad (5.2)$$

where r is the fraction of time the switch is on and in this example, it is 0.8. Thus, the output voltage across the Load resistor would be 80 V in this example. Figure 5.13 shows the output voltage to be about 80 V.

5.3.5   EXAMPLE 5.5: DC-DC CONVERSION: DC MOTOR CONTROL FOR FOUR-QUADRANT OPERATION

Motor characteristics are typically represented by the torque-speed curve of a motor. For example, the torque-speed characteristic of a permanent magnet DC motor is a straight line with a negative slope in the first quadrant. Other different motor types have similar characteristics that reflect the unique characteristic of that motor. While in traditional applications, motors operate in the first quadrant, in vehicle applications motors need to operate in all four quadrants. This is called four-quadrant operation. During forward motion, when the vehicle is being driven forward the vehicle is operating in the first quadrant (+ve Torque, +ve speed). When

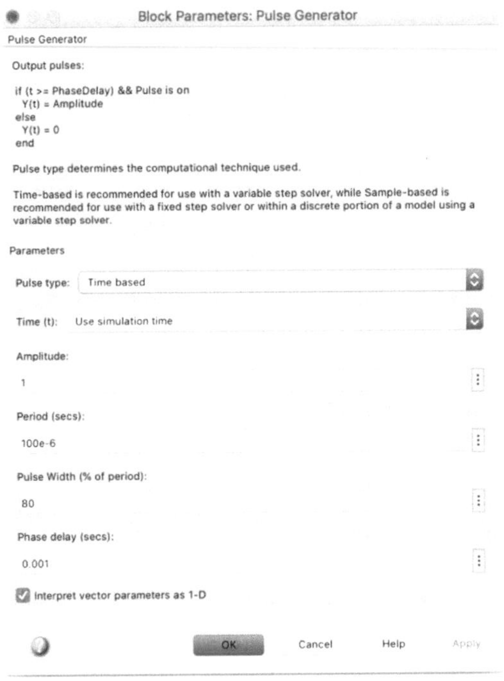

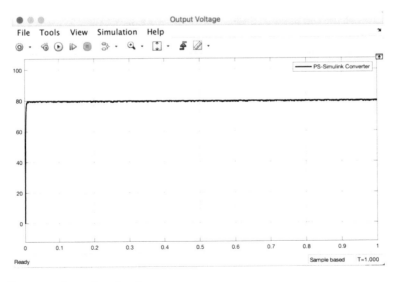

Figure 5.12: Pulse generator setup.

Figure 5.13: Output voltage in a buck convertor.

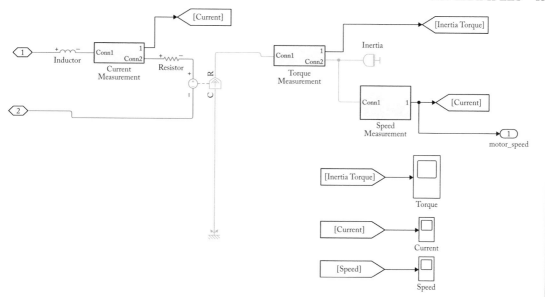

Figure 5.14: DC motor model.

moving in the forward direction but the vehicle is decelerating it is operating in the fourth quadrant (−ve Torque, +ve speed). When the vehicle is speeding up in reverse it is in the second quadrant (+ve Torque, −ve speed) and when the vehicle is braking while in reverse its operation is in the third quadrant (−ve Torque, −ve speed). In EVs and HEVs, during braking the energy recovered from braking is used to charge the battery through the motor or generator. The electric machine works as a generator during this phase. The rest of the time, it operates as a motor. Power electronics and associated controls make this four-quadrant operation smooth. This example demonstrates how this is done with a DC motor.

To build the DC motor four-quadrant operation model we will start with the PMDC Motor model used in Chapter 4.

1. Take the Permanent Magnet DC motor model developed in Chapter 4 and add two input blocks and an output block. The input blocks will replace the motor supply or voltage source and the output block will be called motor_speed and will carry the speed information. The model will look like the picture shown in Figure 5.14.

2. Choose all the items in the model and create a subsystem. Name the subsystem Motor_model.

3. Create an IGBT and Diode circuit combo as shown in Figure 5.15.

4. For the diode setting ensure that the diode is at the default setting as shown in Figure 5.16.

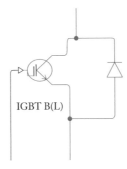

Figure 5.15: IGBT-diode combo.

Figure 5.16: Diode settings.

5. Duplicate the IGBT_Diode combo three more times and add them to the model.

6. Then connect the elements created so far along with a DC voltage source as shown in Figure 5.17. Set the Voltage source value to 15 V. Also add four from blocks and name two of them IGBT_Forward and the other two IGBT_reverse, as shown in Figure 5.17. Also, name the IGBT_diode combos IGBT A(H), IGBT A(L), IGBT B(H), and IGBT B(L), as shown in Figure 5.17. Those represent the IGBTs controlling the motor terminals A and B with high and low settings.

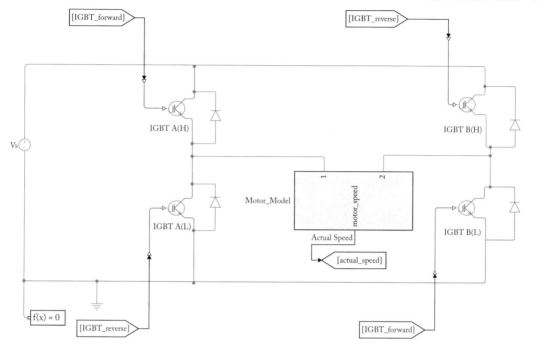

Figure 5.17: Motor subsystem with 4-quadrant connections.

7. We will now build the desired speed input to this model. To test the DC model for conditions where the model can switch direction of motion even though it is driven by a single DC source as well as accelerate and decelerate, we create a speed input profile. The profile is made of a combination of sinusoid as well as a pulse input. The sinusoid function of Amplitude 1 unit and frequency 10 Hz is used, as shown in Figure 5.18. Superposed with that is a pulse input of amplitude 1 unit and period of 1 s (therefore, frequency of 1 Hz) with a 25% cycle on-time, as shown in Figure 5.19. The two inputs are added to form the total input waveform, as shown in Figures 5.20 and 5.21.

8. Create a new subsystem and call it Control. We will now build the controller that will provide the appropriate control signal to the DC motor so that the IGBTs can help switch the direction of the power supply for the motor as well as supply the appropriate level of power to drive the motor at the right speed.

9. Open the Control subsystem and add the following elements in it: two inport and two outport blocks, two rate transition blocks, two addition subtraction block, a PI Controller, and a PWM generation subsystem. Connect them, as shown in Figure 5.22. For the PI parameters set the P parameter to 1 and the I parameter to 5, as shown in Figure 5.23.

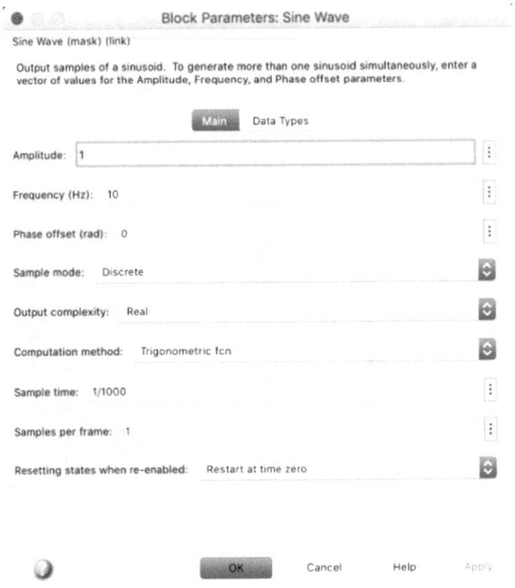

Figure 5.18: Sinusoidal input.

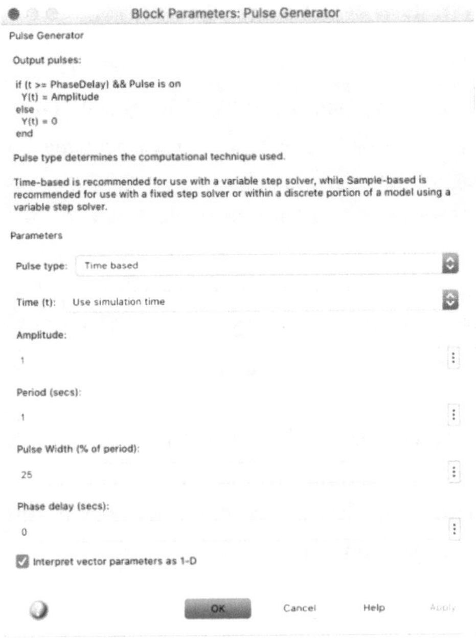

Figure 5.19: Pulse input.

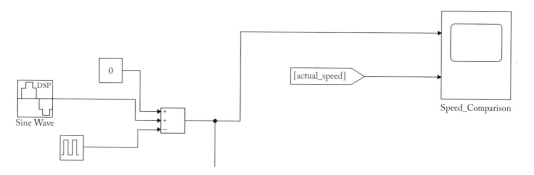

Figure 5.20: **Input speed.**

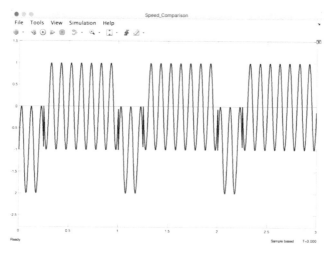

Figure 5.21: **Input speed waveform.**

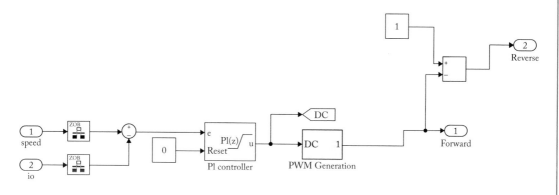

Figure 5.22: **Control subsystem.**

Figure 5.23: PI controller settings.

The input received by the PI controller is the error signal that is the difference between the desired speed and the actual speed.

10. The output of the controller is fed into the PWM subsystem. This is an important component of the control subsystem. As may be recalled, the source voltage used for the PMDC motor is a constant voltage supply of 15 V. But the voltage to be supplied to the motor needs to be a variable DC. Using a PWM signal to keep the motor circuit on for different duty cycles can achieve this goal. The PI signal fed into the PWM subsystem helps generate the appropriate PWM signal. The PWM subsystem is shown in Figure 5.24. The key component of this subsystem is a PWM generation block which receives the desired DC input from the controller and generates an appropriate PWM signal. Figure 5.25 shows the settings of this block and Figure 5.26 shows a sample PWM signal which is a pulse wave of amplitude 1 and varying period that is determined by the controller input.

11. The output of the PWM subsystem is connected to the forward outport and the output subtracted from 1 is connected to the reverse outport. This means when the forward mode is on the reverse mode will remain off and when the reverse mode is on the forward mode will remain off.

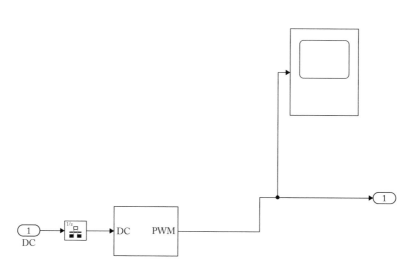

Figure 5.24: **PWM** generation subsystem.

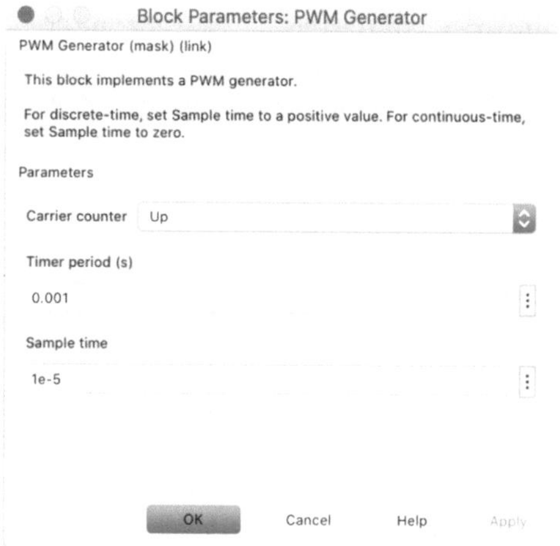

Figure 5.25: **PWM** generator settings.

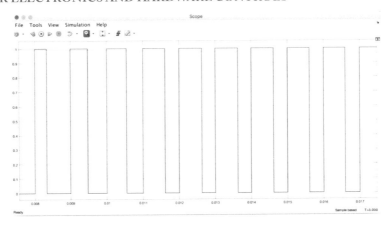

Figure 5.26: Sample PWM signal.

12. The entire model is shown in Figure 5.27. The control subsystem outputs are fed into two Goto blocks named IGBT_forward and IGBT_reverse, respectively. The From blocks of this same name are used to turn pairs of IGBTs on and off to control the motor motion. Thus, the IGBTs A(H) and B(L) operate in unison and the IGBTs A(L) and B(H) operate in unison and the current paths are in one direction in one case and in the opposite direction in the other, resulting in forward and reverse motions.

13. Figure 5.28 shows a comparison of the desired and actual speed of the motor in 4-quadrant operation. With the speed profile used we have demonstrated all four motions, forward movement with increasing speed, forward movement with decreasing speed, reverse movement with increasing speed, and reverse movement with decreasing speed.

5.3.6   EXAMPLE 5.6: DC-AC CONVERSION: 3-PHASE INVERTER

Figure 5.29 shows the inverter circuit that uses six IGBTs to convert a DC input voltage to a 3-phase AC output voltage. This figure shows a circuit which can be found in many references. The load consists of three resistors shown in the model as resistors A, B, and C. For a balanced 3-phase circuit these resistances are the same. In typical applications the load is usually a resistor-inductor combination (coils of motors, etc). But to keep things simple we are using only resistive loads for this example. In the model there are three subsystems connected in parallel to load resistors. These are measurement sensor subsystems for measuring voltage across each of these resistances. The content of one of the subsystems is shown in Figure 5.30. Like in many of the previous models discussed in this text, Goto and From blocks are used to collect and retrieve data for plotting (the three phase voltage). The supply voltage is 200 V per battery, i.e., 400 V. Each load resistor is 1 Ohm. Figure 5.31 shows the block parameters of each IGBT electronic element.

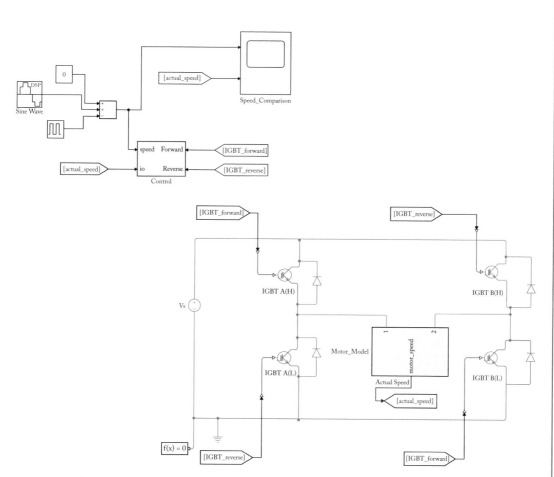

Figure 5.27: The entire model.

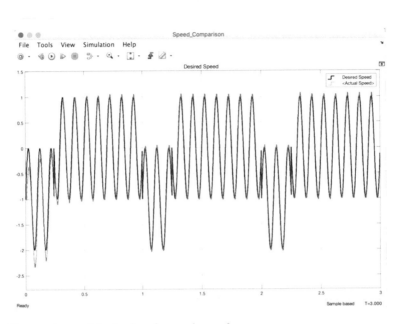

Figure 5.28: Comparison of desired and actual speeds.

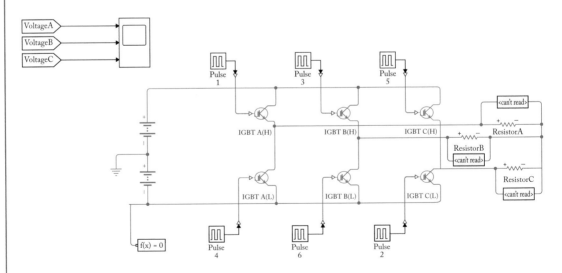

Figure 5.29: Inverter model.

Figure 5.30: Voltage measurement submodel.

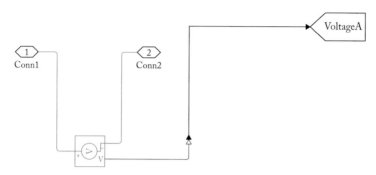

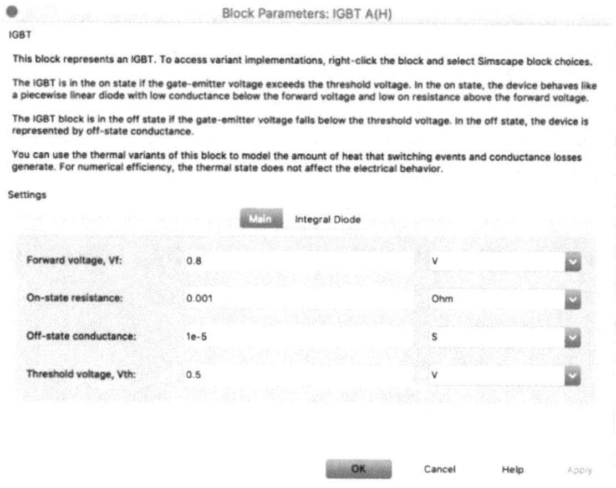

Figure 5.31: IGBT parameters.

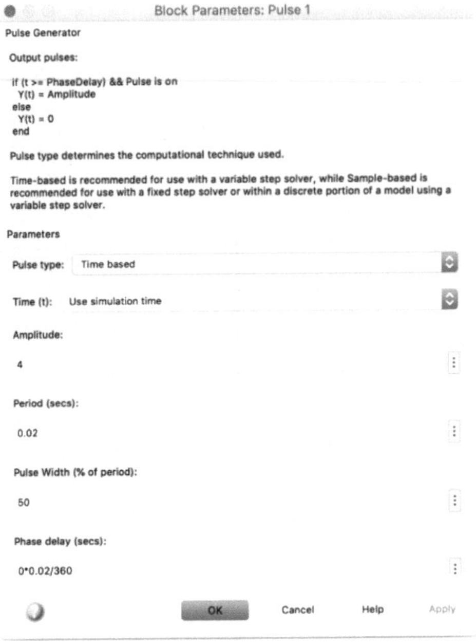

Figure 5.32: Pulse 1 used for first IGBT.

Once again, the switching frequencies of the six IGBTs and their phase differences make the circuit work in the desired fashion. Figure 5.32 shows the setup for pulse 1 which is applied to IGBT A. The pulse amplitude is 4 (this should be nominally greater than the threshold voltage of the IGBT (which is set at 0.5) in the previous figure. The period is set at 0.02 s. This means the frequency is 1/0.02 = 50 Hz. The pulse is on for 50% of time and turned off 50% of the time. The six IGBTs are setup the same way but set apart by a phase difference. Each one is 60° apart in the following fashion:

Phase of Pulse1 = 0*0.02/360

Phase of Pulse2 = 60*0.02/360

Phase of Pulse3 = 120*0.02/360

Phase of Pulse4 = 180*0.02/360

Phase of Pulse5 = 240*0.02/360

Phase of Pulse6 = 300*0.02/360

The simulation is run for 2 s and the output voltages are shown in Figure 5.33 as a 3-phase AC output.

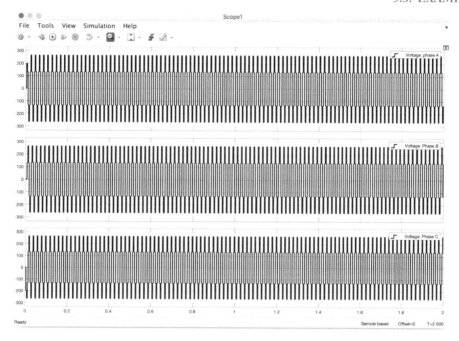

Figure 5.33: Three-phases of the alternating voltage output.

5.3.7   EXAMPLE 5.7: DC-AC CONVERSION: 3-PHASE INVERTER WITH PWM CONTROL

The more commonly used method for PWM controlled 3-phase inversion follows a slightly different approach. In the sinusoidal PWM operation a sinusoidal signal v_A is compared with a high frequency triangular waveform V_T to generate the inverter switch signal. So, switch is on when the sinusoid is greater than the triangle but off when it is smaller than the triangle. This is done for all three phases. The resulting signal modulates the switch duty ratio and its frequency. The frequency of VT establishes the switching frequency. The frequency of the sinusoid signal establishes the output voltage frequency. And the six switches are turned on and off using the following logic. The instructions for building the model and the model itself is listed below.

- S1 on when $V_A > V_T$

- S5 on when $V_C > V_T$

- S3 on when $V_B > V_T$

- S2 on when $V_A < V_T$

- S6 on when $V_C < V_T$

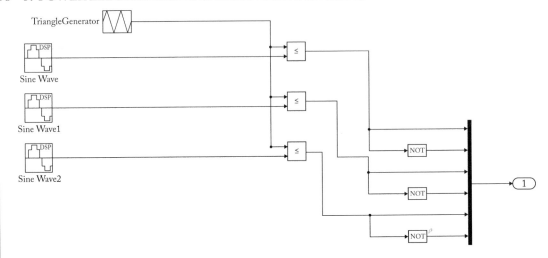

Figure 5.34: PWM control logic.

- S4 on when $V_B < V_T$

1. We will start by establishing the controller. Create a blank subsystem and name it PWM. Just like it is described in the above description the PWM controller is built by including three sine waves and a triangular wave. The sine functions all have a frequency 100 Hz and amplitude of 1 unit. They are phased apart from each other by 120°. The triangular wave also has a peak amplitude of 1 and a frequency of 10,000 Hz (a frequency that is much higher than the sinusoidal waves). Three less than equal to blocks and three NOT blocks are also included. And a mux block is added so that the six outputs can be bunched together. The content of the subsystem is put together, as shown in Figure 5.34.

2. The output of the PWM subsystem is a logical output. Add a demux block to separate the six sub-outputs and then add a double block to convert the data type to double. And store each of the six values into six Goto blocks named S1–S6. These are the six switches that will turn the IGBTs on and off. Figure 5.35 shows the controller subsystem.

3. The rest of the model is built in three parts or three subsystems. They will be the inverter circuit, the load circuit, and the filter circuit. Add three subsystems and name them inverter, filter, and load. Many parts of this model will be the same as the previous example. The load part of the previous example, i.e., three load resistances and the three associated line voltage measuring subsystems will remain identical. Figure 5.36 shows this part of the model. Each resistor value is set at 1 Ohm.

4. The invertor subsystem circuit is shown in Figure 5.37. This is similar to the inverter circuit from the previous example. S1–S6 From blocks are used to represent the six switches.

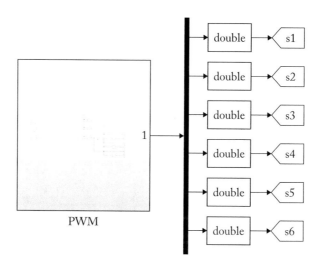

Figure 5.35: PWM controller subsystem.

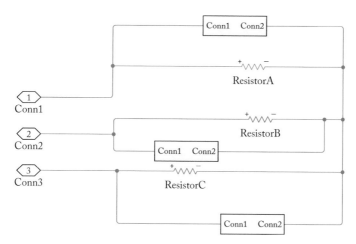

Figure 5.36: Load circuit.

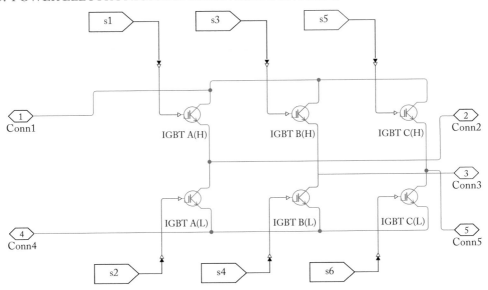

Figure 5.37: Inverter circuit.

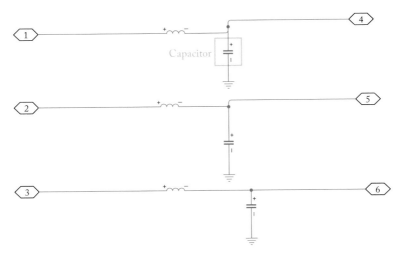

Figure 5.38: Filter circuit.

5. The filter subsystem is shown in Figure 5.38. The inductor for each phase is set at 1e-3 H and the Capacitor value is set at 25e-6F.

6. The three subsystems are connected in a circuit along with two battery packs, as shown in Figure 5.38. Each battery voltage is set to 200 V.

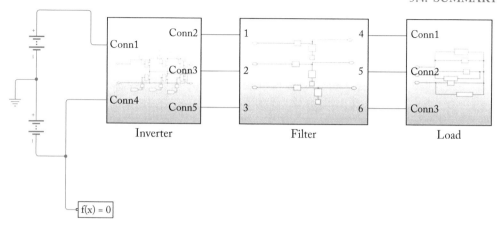

Figure 5.39: Inverter, filter, and load with power supply.

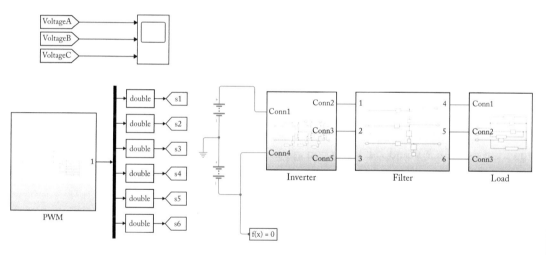

Figure 5.40: Entire model.

7. Once assembled the entire model is shown in Figure 5.40. And the Simulation results showing 3-phase voltages are shown in Figure 5.41. It is clear from the plot how the input supply from a DC source is separated into a 3-phase AC output.

5.4   SUMMARY

In this chapter on modeling various power electronic devices, we covered all the key devices that play important roles in EV and HEV power electronics applications. Attempt was made here to explore the models of actual hardware to demonstrate how certain objectives are achieved. In the

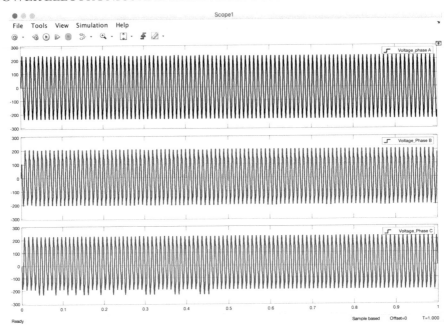

Figure 5.41: Model output.

next chapter we will look into how some of these devices are used in controlling AC machines which are used for EV and HEV traction applications.

CHAPTER 6

Modeling and Control of AC Motors for Traction Applications

6.1 INTRODUCTION

In Chapter 4, we discussed models of two DC motors—the Permanent Magnet DC motor and the separately excited DC motor. Both motors are simple in design and construction and could have been possible candidates for traction applications. However, both these motors are not quite suitable due to a few significant drawbacks. The heat generated in the rotating armature is hard to dissipate for anything more than small motors used for lower power applications. This is because increasing the armature current (a means of controlling these motors) will eventually generate higher heat and dissipating that heat becomes a challenge. Also, use of carbon brushes and commutator rings to switch the current direction leads to the possibility of wear and tear and, therefore, frequent maintenance needs which is not quite acceptable for vehicular applications. A separately excited DC motor addresses part of the problem with the PMDC motor. There are two ways to control the separately excited DC motor, by changing the armature current and by changing the field current. This way, armature overheating can be kept in control. The problem of brushes and commutator rings (and associated maintenance needs) remains the same in the separately excited DC motors as it is in the case of the PMDC. In the end, neither of these are used in the traction applications for HEVs and EVs. A permanent magnet synchronous machine (PMSM), an AC machine, is the most commonly used traction machine today. Of all the motors available today the PMSM machines have the best power density, they can generate the highest amount of power per unit motor mass. Figure 6.1 shows a cut-out view of the motor drive of a Ford Escape Hybrid vehicle which uses two electric machines, a motor and a generator. Both are PMSM machines and are identical in construction. In Figure 6.1 the stator of one of the machines can be clearly seen as well as the rotor and stator of the second machine.

In the Permanent Magnet Synchronous Machine (PMSM) the rotor is a permanent magnet. Figure 6.2 shows two views of a PMSM rotor. The first view shows how the magnets are arranged close to the surface of the rotor. With the rotor being a permanent magnet the heat generation problem in the rotor becomes a non-issue for the PMSM motors. The stator for the PMSM machine is an electromagnet or a coil of wire wound on a laminated core. In fact, it is

Figure 6.1: Ford Escape Hybrid electric drive.

Figure 6.2: Two views of the PMSM rotor.

three separate coils of wire that are electrically connected in the standard Y-connection that is common for AC systems. Figure 6.3 shows a picture with different views of the stator from the Ford Escape Hybrid drive.

The three stator coils (a,b,c) supply three currents ia, ib, and ic in a manner that the currents are not only phase-separated in time (by 120°) but they are also phase separated in space. For example, Figure 6.4 shows a schematic representing the distribution in space of phase a. The

Figure 6.3: Two views of the PMSM stator.

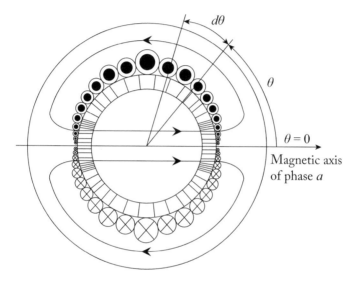

Figure 6.4: Schematic showing phase a of coils on the stator.

coils are arranged such that the phase centerline is in the horizontal x direction. Phase *b* is 120° counterclockwise from the horizontal x-direction. And phase *c* is 240° counterclockwise from the horizontal x-direction. As a result of this arrangement the three coils generate three magnetic fields that are spaced apart 120° in space. Also, because of the time separation of the three currents, the magnetic field from each phase is separated by 120° in time. As a result of these two types of phase separations, the combination of the three magnetic fields is a resultant magnetic field that rotates, even though the magnetic fields of each phase is stationary. The speed of rotation of the resultant magnetic field is the same as the current frequency, hence the name synchronous.

In the PMSM machine, therefore, there is one rotating magnetic field generated by the stator coils, and a second rotating magnetic field due to the rotor, which is a permanent magnet. This is a big difference between PMSM machines and DC motors where despite the physical rotation of the rotor the two magnetic fields are stationary at all times and are 90° apart from

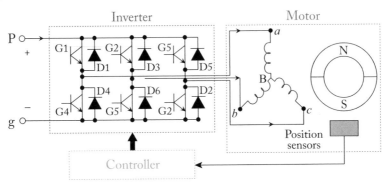

Figure 6.5: Simple schematic of a PMSM drive.

each other. In order for any motor to work properly and at maximum efficiency, the two magnetic fields (stationary or rotating) need to be 90° apart from each other.

Keeping the two magnetic fields 90° from each other is therefore an additional control task for PMSM machines. To do this it is necessary to know where the rotor is at any given point in time. PMSM machines use sensors (and recently, sensor-less techniques) to continuously track the location of the rotor magnetic field and uses that information in the control algorithm to determine the size and location of the stator magnetic field (90° away from the rotor magnetic field). Figure 6.5 shows a simple schematic for the operation of the PMSM machine drive. The sensor senses the location of the rotor magnetic field and sends that information to the controller. The controller determines the 3-phase currents necessary to generate the appropriate stator magnetic field (magnitude and location) and uses PWM controller to drive the switches of the inverter to drive the appropriate amounts of current in each phase. The source of power is the battery which is a DC source and the PWM controller along with the power electronic devices such as the invertor generates AC power to supply the motor.

In this machine, the rotor inside the motor is a permanent magnet while the stator is an electromagnet, has three coils and is excited by a 3-phase AC source. The AC source is suitably controlled to replicate the two-pronged control mentioned earlier, current control as well as the field control. The operational envelope that was introduced in Chapter 4 for the separately excited DC motor (as shown in Figure 4.20) is the same envelope that is used for the PMSM traction machines.

6.2   d-q CONTROL OR VECTOR CONTROL

Control of motors is typically a two-pronged process; they are typically referred to as torque control (or current control) and field control. In the case of separately excited DC motors, torque or current control is achieved by increasing the armature voltage/current. There is a limit to how high the armature current can be. This limit is the torque/current limit for the armature.

When this limit is reached, the armature current is held steady and the field current is reduced to weaken the magnetic field. This same control approach is implemented for the PMSM. The implementation is caried out through a process called Vector control or d-q control. The core of this algorithm is a set of mathematical transformations often called d-q transformation or Park's transformation. Through this transformation, the three phase currents (ia, ib, and ic) are transformed into two phases that are perpendicular to each other. One of these directions is called the "direct" or d, which is in-line with the rotor magnetic field. The second one is called the "quadrature" or q, which is perpendicular or 90° ahead of the d direction. Equations (6.1) and (6.2) show the mathematical transformation from a-b-c to d-q and from d-q to a-b-c, respectively. The only information that is needed for this transformation is the angular orientation of the rotor. That is obtained from the sensor that tracks the rotor angular location. In the equation it shows up as the angle θ_{da}.

$$
\begin{bmatrix} i_{sd}(t) \\ i_{sq}(t) \end{bmatrix} = \sqrt{\frac{2}{3}} \underbrace{\begin{bmatrix} \cos(\theta_{da}) & \cos\left(\theta_{da} - \frac{2p}{3}\right) & \cos\left(\theta_{da} - \frac{4p}{3}\right) \\ -\sin(\theta_{da}) & -\sin\left(\theta_{da} - \frac{2p}{3}\right) & -\sin\left(\theta_{da} - \frac{4p}{3}\right) \end{bmatrix}}_{[T_s]_{abc \to dq}} \begin{bmatrix} i_a(t) \\ i_b(t) \\ i_c(t) \end{bmatrix} \tag{6.1}
$$

$$
\begin{bmatrix} i_a(t) \\ i_b(t) \\ i_c(t) \end{bmatrix} = \sqrt{\frac{2}{3}} \begin{bmatrix} \cos(\theta_{da}) & -\sin(\theta_{da}) \\ \cos\left(\theta_{da} + \frac{4\pi}{3}\right) & -\sin\left(\theta_{da} + \frac{4\pi}{3}\right) \\ \cos\left(\theta_{da} + \frac{2\pi}{3}\right) & -\sin\left(\theta_{da} + \frac{2\pi}{3}\right) \end{bmatrix} \begin{bmatrix} i_{sd} \\ i_{sq} \end{bmatrix}. \tag{6.2}
$$

Transforming the currents from three phase to two phase allows us to implement the two-pronged control. By holding the d current steady and increasing the q current, it allows us to implement the torque control. When the torque has reached a pre-defined limit, it can be held steady and the d current can be altered (in fact reduced) to implement flux weakening. The d direction already has a steady magnetic field from the permanent magnet. So, a negative d current helps create a magnetic field opposite to that magnetic field from the permanent magnet and effectively weakens the existing field. For more details on the operation and control of PMSM machines the reader should refer to texts on this subject.

6.3   EXAMPLES

Following are a handful of tips to add new elements into a model.

Tips for adding model elements

1. Use Quick Insert to add the blocks. Click in the diagram and type the name of the block. A list of blocks will appear and you can select the block you want from the list. Alternatively, the Open Simscape Library block can be used to look though the library of all blocks and pick the appropriate one.

2. After the block is entered, a prompt will appear for you to enter the parameter. Enter the variable names as shown below.

3. To rotate a block or flip blocks, right-click on the block and select Flip Block or Rotate block from the Rotate and Flip menu.

4. To show the parameter below the block name, see Set Block Annotation Properties in the Simulink documentation.

6.3.1   EXAMPLE 6.1: MODELING AC MOTORS USING A SIMPLIFIED PMSM DRIVE

Simscape offers a variety of motor model blocks that can be used to model many different types of DC and AC Motors. Since the permanent magnet Synchronous Machine has been the motor of choice for HEV and EV applications we will confine our discussion here to only the PMSM motor. There are two blocks available within the Simscape library that can be used to model the behavior of a PMSM motor. Figure 6.6 shows the two PMSM DC Motor blocks within its libraries. The PMSM block models the PMSM motor from first principles that is, the governing equations are modeled entirely. This block has to be used in conjunction with appropriate controllers and power electronics. Also, this is driven by 3-phase AC supply. We will demonstrate an example with this block in this chapter. But the first example we have included here uses the Simplified PMSM drive block. As the name suggests, it is a simplified block that is good for system level modeling. It uses a DC input and is designed to provide torque and speed outputs that are typical of a PMSM motor. The block uses a steady state torque speed relationship curve to determine the motor output (instead of the fundamental electromagnetic equations). So, implementing a system model using this motor is easier. Also, elaborate vector controllers, AC source, and power electronics blocks are not necessary when using this block.

Figure 6.7 shows a typical Torque speed envelope for a PMSM machine. The two distinct regions are the constant torque region and the constant power region. The motor is controlled using armature current in the former region and through flux-weakening in the latter. For the Simplified PMSM drive block data similar to this plot is used to implicitly control the motor output instead of modeling the behavior of the motor from first principles.

In this example we will demonstrate how to use the PMSM block to model a PMSM drive for a vehicle.

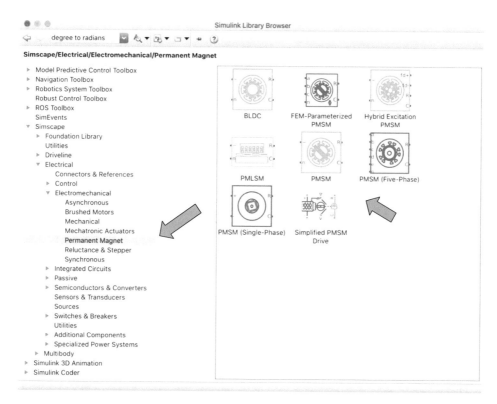

Figure 6.6: Locating the PMSM motor block in the electrical library.

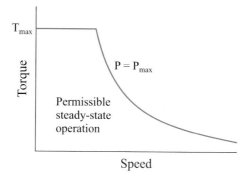

Figure 6.7: Typical torque speed envelope for a PMSM motor.

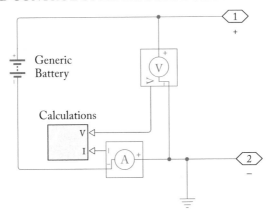

Figure 6.8: Building the battery model.

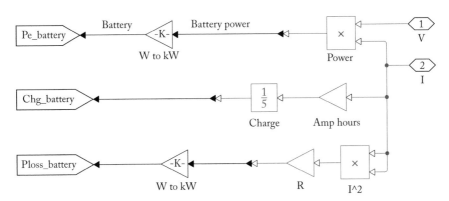

Figure 6.9: Subsystem calculations.

Steps:

1. Type **ssc_new** in the Matlab command window to open a new model file.

2. From Simscape library add a Generic battery, a Voltage Sensor, a Current sensor, and an Electrical reference (ground). Also add two PMC ports. Connect them, as shown in Figure 6.8. Add a subsystem and name it Calculations.

3. Open the subsystem and Create the calculation system shown in Figure 6.9 to calculate the state of charge, battery losses, and the power drawn from the battery. Figures 6.10 and 6.11 show the Generic Battery block settings used in this example.

4. In the Calculations subsystem the product of current and voltage calculates the total power expended and the Amp hours calculates the charge that is used. An initial condition of

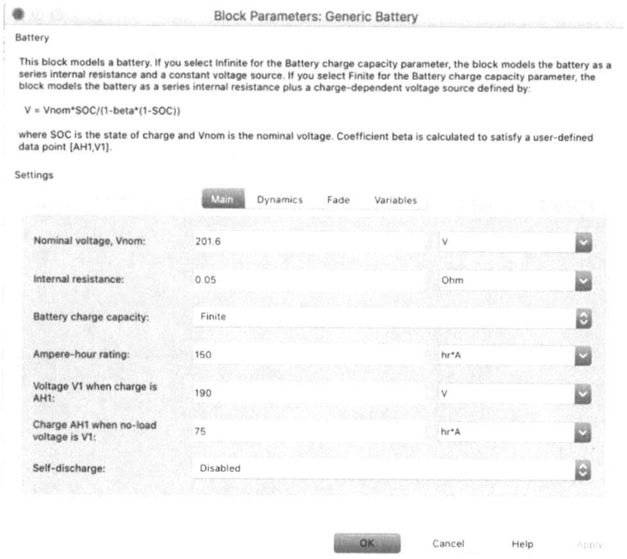

Figure 6.10: Generic battery block settings used.

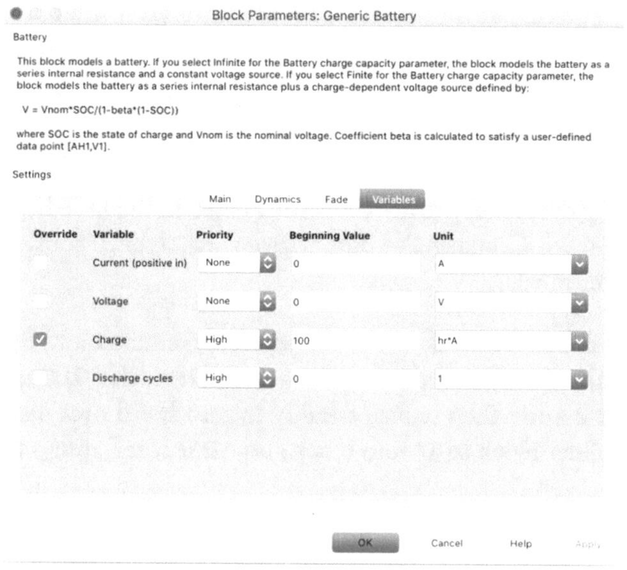

Figure 6.11: Generic battery variables settings.

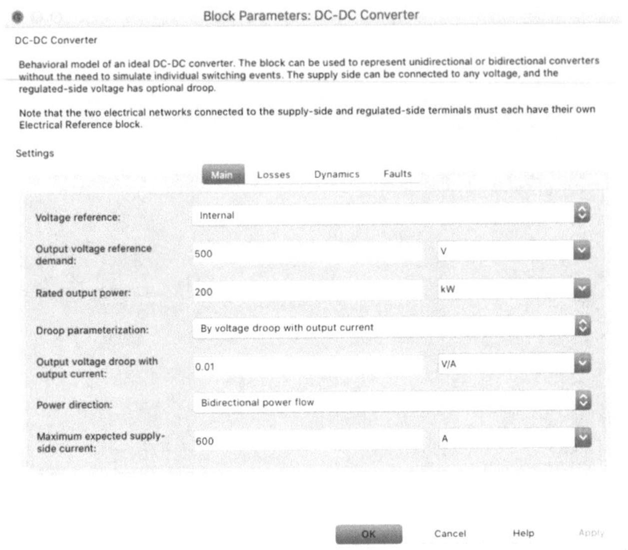

Figure 6.12: DC-DC convertor block settings.

150 Ahr is used as the initial charge in the battery. The used charge is subtracted from the initial value to calculate the battery state of charge. And the current square times the battery resistance calculates the power loss in the battery. Store all the calculations in Goto blocks called, Pe_battery, Chg_battery, and Ploss_battery, respectively (Figure 6.9).

5. Choose everything on the screen so far and create a subsystem. Name it Battery.

6. Add a DC-DC convertor block to the model. This block models a buck-boost convertor without modeling the actual circuit. The settings used in the DC-DC convertor are shown in Figures 6.12–6.15. Connect the output from the Battery subsystem to the DC-DC convertor inlet terminals.

7. Now add a Signal builder block to the model from Simulink and build the signal shown in Figure 6.16. The motor will speed up for 4 s until it reaches 100 rpm. It will stay steady at that speed for 2 s and then reduce steadily to zero speed over 4 s. Connect the output of the Signal Builder block to a Goto block and name it ref_rpm_motor.

8. The assembled model so far should look something like Figure 6.17.

9. We will now create the motor. Add subsystem and name it Motor. Open this subsystem and add the Simplified PMSM Drive, a Mechanical rotational reference, a Torque sensor, four subsystems, and three PMC ports.

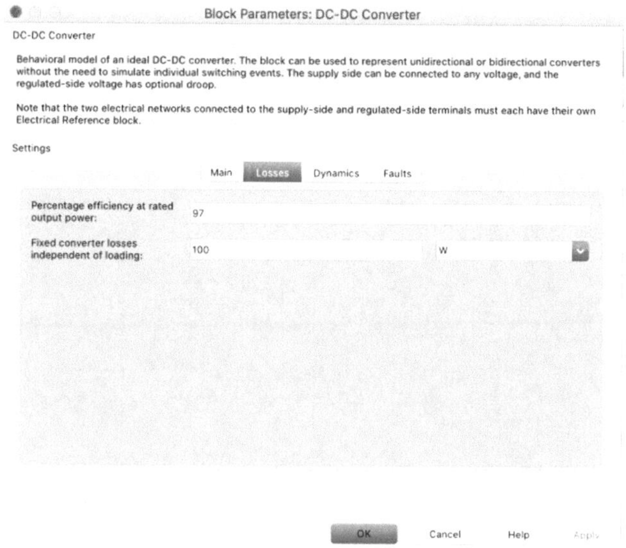

Figure 6.13: DC-DC convertor losses.

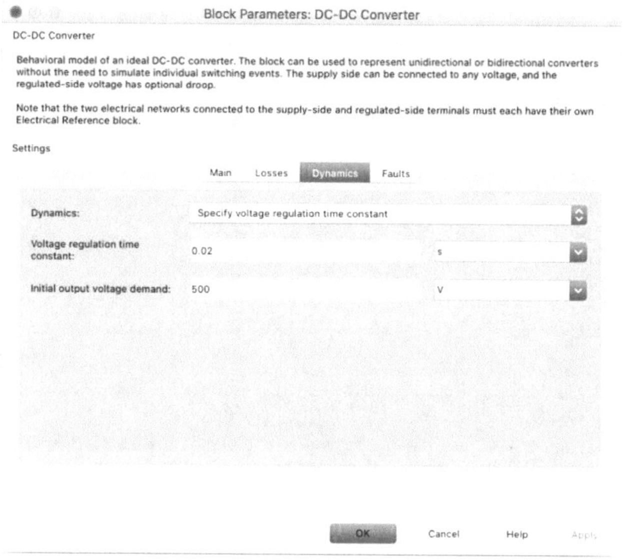

Figure 6.14: DC-DC convertor dynamics.

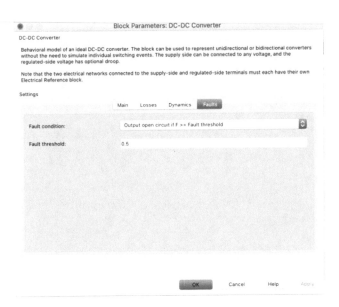

Figure 6.15: DC-DC convertor faults.

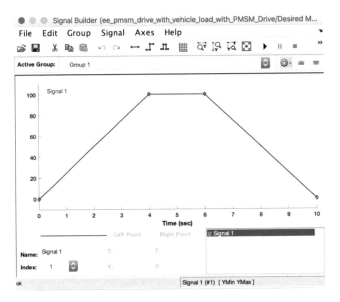

Figure 6.16: Demanded motor speed.

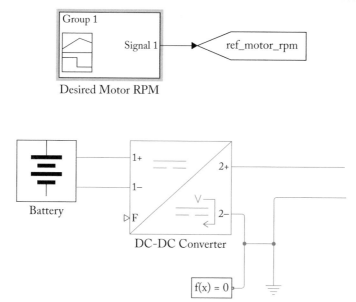

Figure 6.17: A partial model.

10. Name the first subsystem Electrical sensors and open it. Add a current sensor and a voltage sensor and 3 PMC ports. Include two PS-Simulink blocks. Set one block's unit to A and the other one to V. Connect these two blocks to two Goto blocks. Name the two Goto blocks motor_current and motor_voltage, respectively. Connect everything, as shown in Figure 6.18.

11. Name the second subsystem Mechanical Power and open it. Add two PMC ports; name them w and trq. These will bring in angular velocity and torque information. Add a PS product block a Ps_Simulink block and a gain block. Set the gain value to 1/1000/. Connect them, as shown in Figure 6.19 and connect the output of the gain block to a Goto block named Pm_motor.

12. Name the third subsystem RPM sensor and open it. Add a rotational motion sensor, a rotational reference block and a PS_Simulink block. Also add a PMC port and an outport. Set the unit in the PS_Simulink block to rpm. And create the connections, shown in Figure 6.20. Add a GoTo block, name it motor_rpm and connect, as shown in Figure 6.20.

13. Name the fourth subsystem Speed Controller subsystem. We discussed earlier that the Simplified PMSM Drive uses a Torque input and provides a desired speed as the output. This torque will be called reference torque. This block is essentially a PI controller which uses the error in the speed of the motor to generate the torque signal that will be used as an input to the motor. For a 2-level cascade controller (discussed in Chapter 4) this

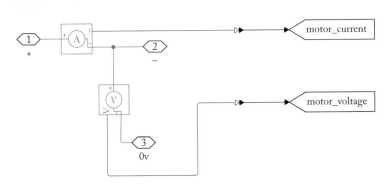

Figure 6.18: Subsystem electrical sensors.

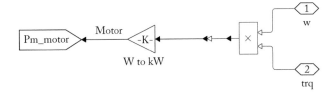

Figure 6.19: Mechanical power subsystem.

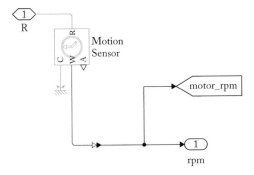

Figure 6.20: RPM sensor subsystem.

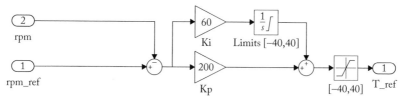

Figure 6.21: Speed controller subsystem.

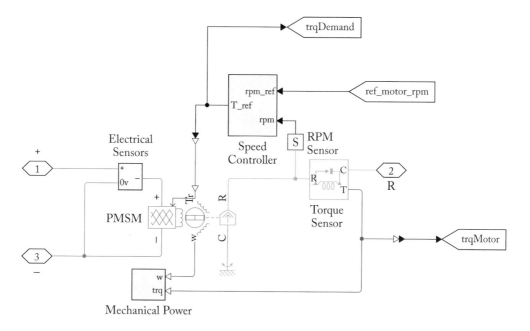

Figure 6.22: Motor subsystem.

controller is essentially the inner loop of a cascaded controller. Open Speed Controller subsystem and add two inports and one outport, two addition-subtraction blocks, two gains for proportional and integral constants, an integrator with limits, and a saturation block. Connect them as shown in Figure 6.21. Set the Ki value to 60 and Kp value to 200. Set the integration limit from −40 to 40 and the saturation limit to ±400. This will mean that the torque will be limited to 400 Nm.

14. Once the four subsystems are created build the Motor subsystem, as shown in Figure 6.22. Make sure the two Ps_Simulink blocks have their units set to Nm. Two Goto blocks are used here to keep track of the motor torque, the torque demand and one From block is used to recall the ref_motor_rpm data.

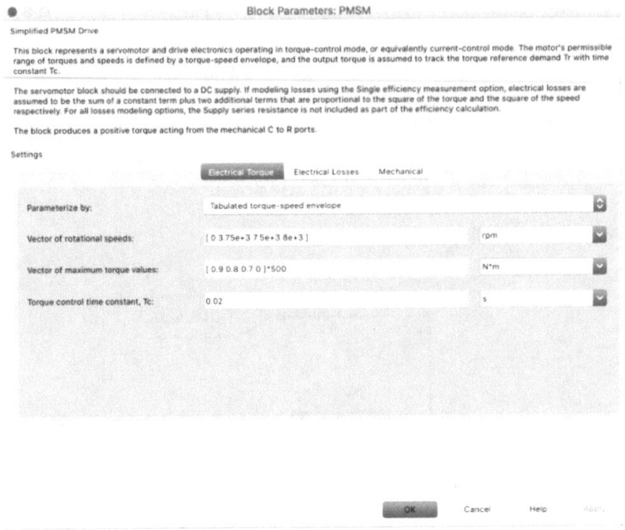

Figure 6.23: Simplified PMSM drive, electrical torque.

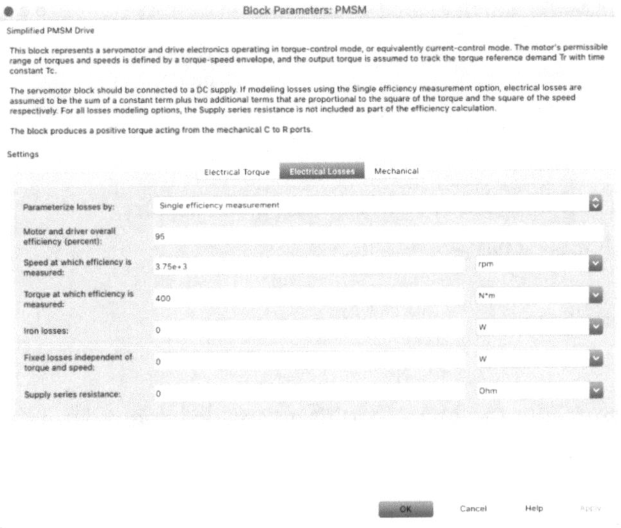

Figure 6.24: Simplified PMSM drive, electrical losses.

15. Figures 6.23, 6.24, and 6.25 show the settings for the PMSM Drive to be used for this model.

Figure 6.25: Simplified PMSM drive, mechanical.

16. We will now add the load/resistance force for the vehicle. In other words, the vehicle model has to be added. The net traction force F (as shown in Equation (6.3)) applied from the motor drive is used to overcome the three dominant sources of resistance, the road frictional resistance, gravitational resistance if the car is moving up a slope, and the wind resistance due to wind drag. The net force is then utilized to accelerate the vehicle.

$$ma = F - \mu mg - mg\,Sin\theta - \frac{1}{2}\rho C_d\,Av^2. \tag{6.3}$$

17. Add a subsystem and call it Vehicle Body. Open the subsystem and add three more subsystems in it. Name them Gravity, Road, and Drag, respectively. Also add a wheel and axel block, a liner motion sensor, a Mass, a Mechanical Translational Reference, a Force source, and two PS add blocks. Also add an Inport, an Outport, and a PMC Block.

18. Figures 6.26, 6.27, and 6.28 show the three subsystems that model the three resistance forces from wind drag, road resistance, and gravity, respectively. The mass of the vehicle used is 1200 kg, the coefficient of rolling friction as 0.01, drag coefficient is 0.3, vehicle frontal area of 1.4 sq/m, and air density of 1.225 kg/cum. The slope angle is accepted as an input in degrees through the inport. All three forces are computed in their respective subsystem and multiplied with −1 and then added.

19. Figure 6.29 shows the entire vehicle body subsystem. The net resistive force is then applied using a force source to the vehicle mass. It also receives input from the motor drive. The

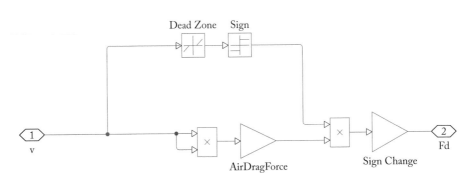

Figure 6.26: Drag subsystem.

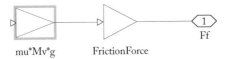

Figure 6.27: Road subsystem.

Figure 6.28: Gravity subsystem.

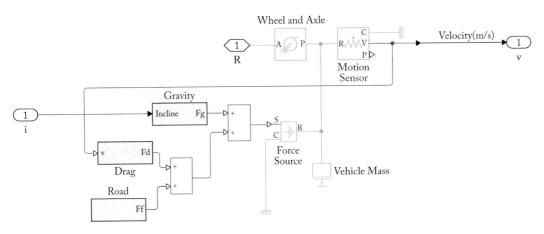

Figure 6.29: Vehicle body subsystem.

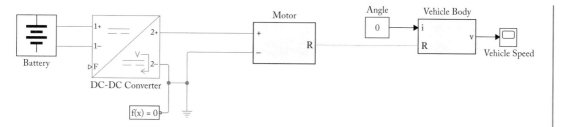

Figure 6.30: Assembled model.

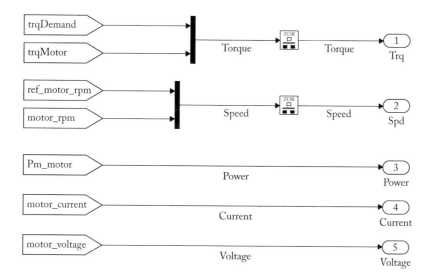

Figure 6.31: Drive signals subsystem.

input torque is converted to a force/thrust using a wheel and axel block to convert torque to force using a wheel radius of 0.3 m. The motion sensor tracks the velocity in m/s and reports it at the Outport. Figure 6.30 shows the assembled model so far.

20. Add two more subsystems to the main model to keep track of all the calculated quantities. Name them Drive Signal and Battery State, respectively. The first one tracks all the quantities associated with the motor drive and the second one tracks the variables associated with the battery. Figure 6.31 shows the Drive Signal subsystem. It is mostly made of From blocks that are associated with Goto blocks that were used throughout the model to store different measured variables. Figure 6.32 shows the Battery state subsystem with the three variables related to battery performance.

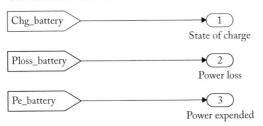

Figure 6.32: Battery state subsystem.

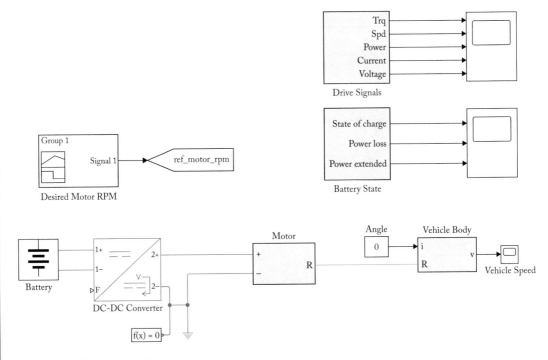

Figure 6.33: Entire model.

21. Connect scopes to all the outputs from the two subsystems. Figure 6.33 shows the entire model after all parts have been created.

22. The simulation is run for 10 s with a 0° input slope for the road. Figures 6.34 and 6.35 show the two output plots.

23. The simulation is run a second time with a 2° input slope for the road. Figures 6.36 and 6.37 show the two output plots.

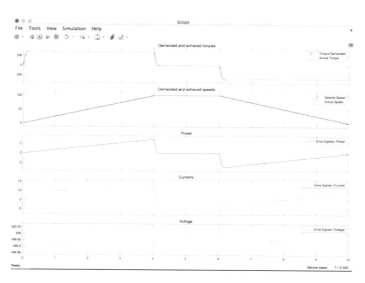

Figure 6.34: The motor variables when the road slope is 0°.

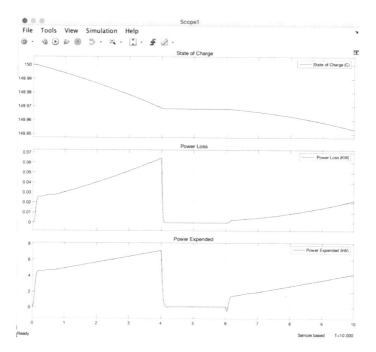

Figure 6.35: The battery variables in the model for a 0° slope.

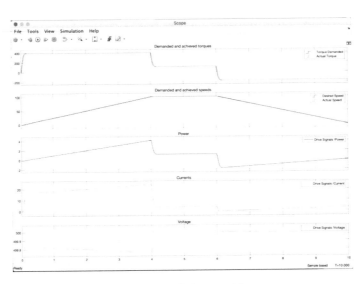

Figure 6.36: The motor variables when the road slope is 2°.

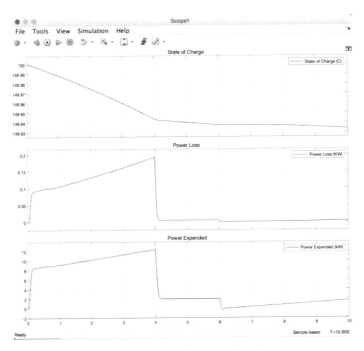

Figure 6.37: The battery variables in the model for a 2° slope.

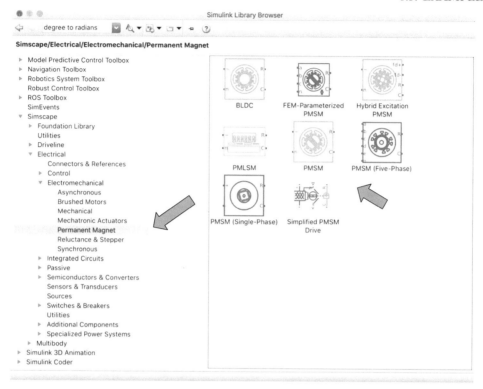

Figure 6.38: Locating the PMSM motor block in the electrical library.

6.3.2   EXAMPLE 6.2: MODELING AC MOTOR AND ITS CONTROL USING A PMSM MOTOR MODEL

As we have discussed before, Simscape offers a variety of motor model blocks that can be used to model many different types of DC and AC Motors. Earlier, we identified two blocks available within the Simscape library that can be used to model the behavior of a PMSM motor. Figure 6.38 shows the two PMSM Motor blocks within its libraries. The PMSM block models the PMSM motor from first principles by modeling the governing equations explicitly. This block has to be used in conjunction with appropriate controllers as well as power electronics. Also, this is driven by 3-phase AC supply. In this example we will demonstrate the use of this block in modeling the Motor drive system with an AC motor. The PMSM motor has two interacting magnetic fields, one generated by the 3-phase power supply supplied by the 3-phase windings in its stator. The second one is the magnetic field of a very strong permanent magnet that is the rotor. In the PMSM motor, both the magnetic fields are rotating and the interaction of the two magnetic fields helps to generate the torque supplied by the motor. This motor is driven by a 3-phase current but is controlled in a manner similar to separately excited DC motors, i.e.,

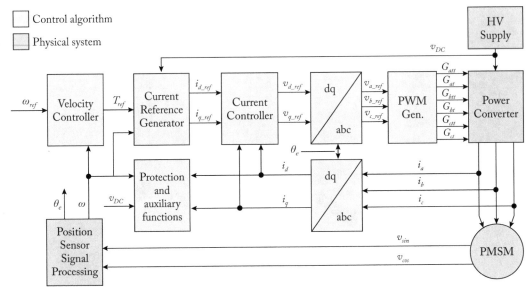

Figure 6.39: PMSM field-oriented control diagram.

both torque/current control as well as field control has to be used to be able to control the motor behavior throughout its operating range. In order to implement this type of control in the AC machine, the d-q control algorithm or vector control is used. In the d-q approach the three phase current is mathematically transformed into a 2-phase current, direct and quadrature. The control algorithm, determines the d and q currents that are to be demanded from the power source and the i_{dq} to i_{abc} mathematical transformation process determines the a-b-c phase currents that are to be demanded from the power source. The power source, coupled with appropriate power-electronics supplies the necessary current to the motor. This fairly complicated control algorithm is implemented through a Simscape block named PMSM Field Oriented Control. Figure 6.39 shows the control algorithm implementation within this block. Using the desired speed and the measured speed information the velocity controller in the outer control loop develops the control signal which is shown in figure as the reference Torque signal. This signal is used to calculate the direct and quadrature reference currents that is compared with the actual direct and quadrature currents and this new error signal is used in the current controller, the inner controller, to generate two control signals for the direct and quadrature circuits. Since the direct and quadrature quantities are only virtual values, they are fed into a dq-abc converter (park transform) which mathematically determines the corresponding three phase voltage demands. Those demands are fed through a PWM generator into the power electronics gates so appropriate supply voltage can be applied to the 3-phase stator circuits. Sensors (real or virtual) measure the speed and location of the rotor as well as the actual three phase currents and all these information is fed back into the control loop at appropriate points in the algorithm. Our use here of PMSM field-oriented control block allows us to perform all of these multi-level control tasks through the use of a single block. We will now describe the development of this model and its key components.

Following are a handful of tips to add new elements into a model.

Tips for adding model elements

1. Use Quick Insert to add the blocks. Click in the diagram and type the name of the block. A list of blocks will appear and you can select the block you want from the list. Alternatively, the Open Simscape Library block can be used to look though the library of all blocks and pick the appropriate one.

2. After the block is entered, a prompt will appear for you to enter the parameter. Enter the variable names as shown below.

3. To rotate a block or flip blocks, right-click on the block and select Flip Block or Rotate block from the Rotate and Flip menu.

4. To show the parameter below the block name, see Set Block Annotation Properties in the Simulink documentation.

Steps:

1. Type **ssc_new** in the Matlab command window to open a new model file.

2. From Simscape library add a Generic battery, and change its setting values as shown in Figures 6.40 and 6.41. Also, choose the battery to not have any dynamic effects. For now, unlike the previous model we will not look at the battery condition (State of Charge, losses, etc.) in this model to minimize the complexity of the model.

3. Add a new subsystem in the model and name it 3-phase inverter. From the examples described in the previous chapter copy and paste the IGBT based inverter into this subsystem. Add three PMC terminals in the subsystem and a Splitter that combines the 3-phase outputs into a single channel. The switching information will be brought in through a Inport named G. Use a Demux block to separate it into six signals for the six gates. After connecting all these components, the subsystem should be like the one shown in Figure 6.42.

4. Connect the battery with a Electrical Reference and the 3-phase Invertor subsystem, as shown in Figure 6.43.

5. Add a new subsystem to the model and call it i. This will be the current measuring subsystem. Open this subsystem and add two PMC blocks and a 3-phase current sensor. Attach them, as shown in Figure 6.44. Also use a Goto block and PS-S block to capture the three phase current data. Connect this subsystem's input to the Inverter output.

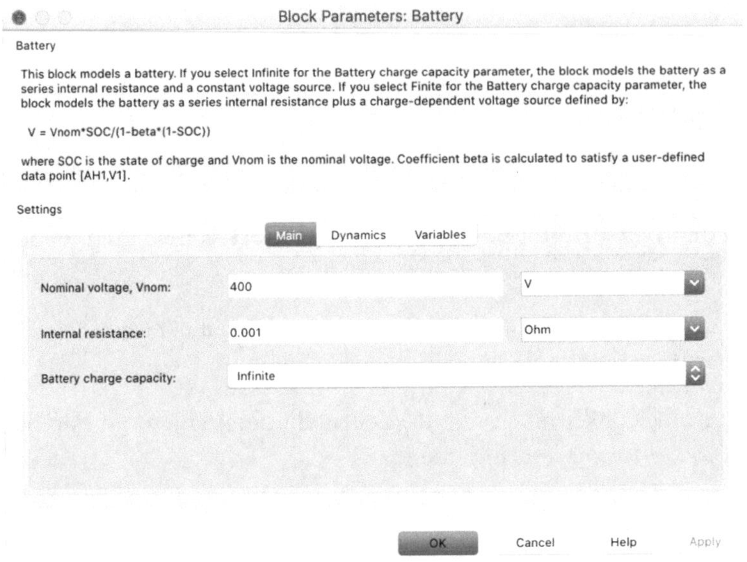

Figure 6.40: Battery supply settings.

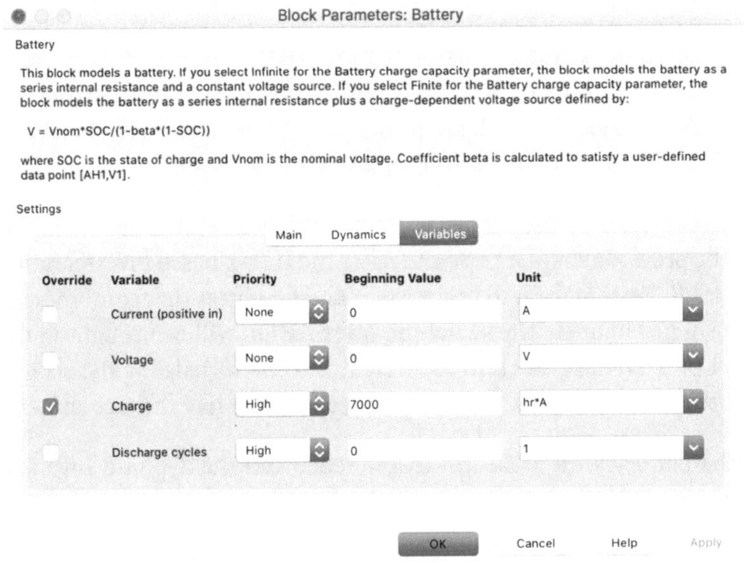

Figure 6.41: Battery variables setting.

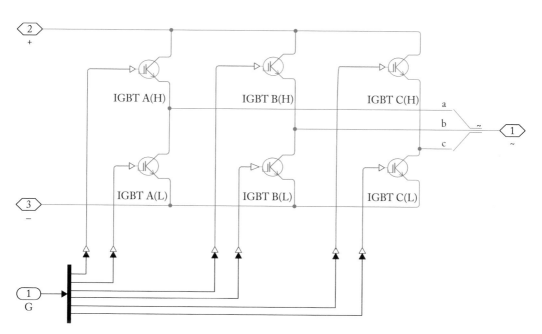

Figure 6.42: 3-phase inverter subsystem.

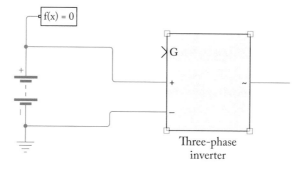

Figure 6.43: Battery and the 3-phase inverter subsystem.

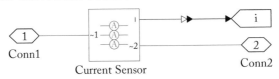

Figure 6.44: Current sensor subsystem.

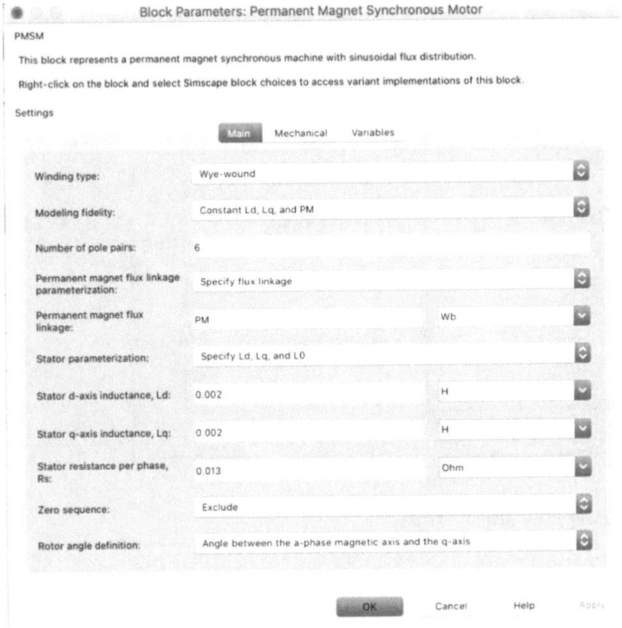

Block Parameters: Permanent Magnet Synchronous Motor

PMSM

This block represents a permanent magnet synchronous machine with sinusoidal flux distribution.

Right-click on the block and select Simscape block choices to access variant implementations of this block.

Settings

	Main	Mechanical	Variables
Winding type:	Wye-wound		
Modeling fidelity:	Constant Ld, Lq, and PM		
Number of pole pairs:	6		
Permanent magnet flux linkage parameterization:	Specify flux linkage		
Permanent magnet flux linkage:	PM	Wb	
Stator parameterization:	Specify Ld, Lq, and L0		
Stator d-axis inductance, Ld:	0.002	H	
Stator q-axis inductance, Lq:	0.002	H	
Stator resistance per phase, Rs:	0.013	Ohm	
Zero sequence:	Exclude		
Rotor angle definition:	Angle between the a-phase magnetic axis and the q-axis		

OK   Cancel   Help   Apply

Figure 6.45: PMSM motor main setting.

6. From the Simscape library add the PMSM motor to the model and set the motor parameters as shown in Figures 6.45, 6.46, and 6.47. Connect the motor electrical input to the current sensor output. Connect the second electrical terminal of the motor to the electrical reference through a large electrical resistor of value 1800 Ohms (Gmin).

7. Add a new subsystem to the model and call it Encoder. This will be used to make mechanical measurements of the motor output. Open the subsystem and add four PMC ports. Also add a Torque sensor and a rotation sensor. Add two Goto Blocks and name them trqmotor and wMotor. Add three PS_S blocks and a Mux Block. Connect them as shown in Figure 6.48.

8. We will now add the mechanical load or the drive load. For this we will add exactly the same load as was used in Example 6.3.1. Copy the Vehicle Body subsystem from the

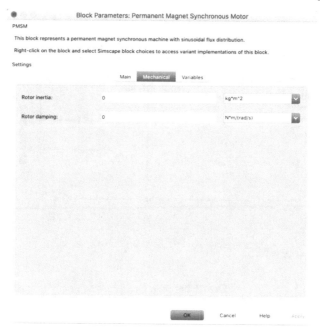

Figure 6.46: **PMSM** motor mechanical setting.

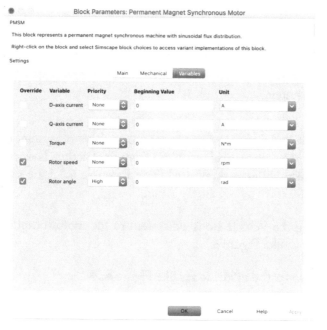

Figure 6.47: **PMSM** motor variables setting.

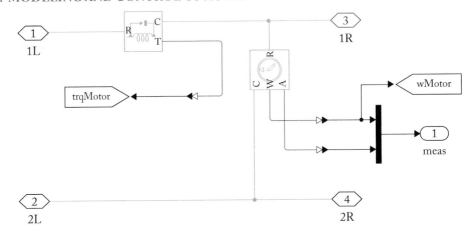

Figure 6.48: Encoder subsystem.

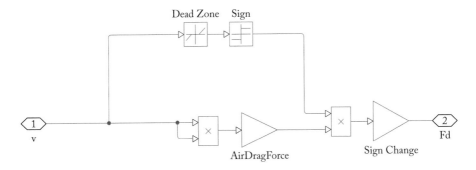

Figure 6.49: Drag subsystem.

previous example and add it to the current model. This subsystem consists of three load types, gravity due to a slope, road resistance from tire friction, and wind resistance or drag. The subsystem assemblies as duplicated from Example 6.3.1 are shown in Figures 6.49, 6.50, and 6.51.

9. After connecting the Vehicle Body subsystem to the motor output the assembled drive system so far looks like Figure 6.52.

10. After assembly so far the model looks like Figure 6.53.

11. We will be using the same motor speed demand as in the previous example. So copy the signal builder (Figure 6.54) from the previous example and the attached Goto block ref_motor_rpm and add that to the current model.

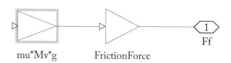

Figure 6.50: Road subsystem.

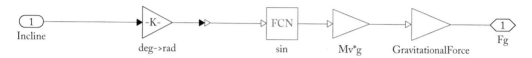

Figure 6.51: Gravity subsystem.

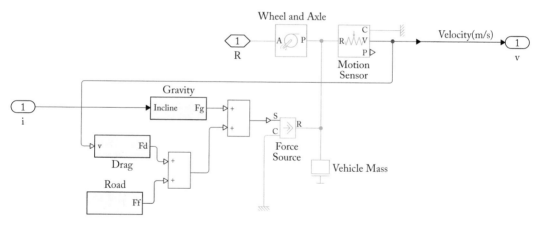

Figure 6.52: Vehicle body subsystem.

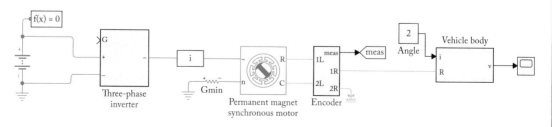

Figure 6.53: Partially assembled model.

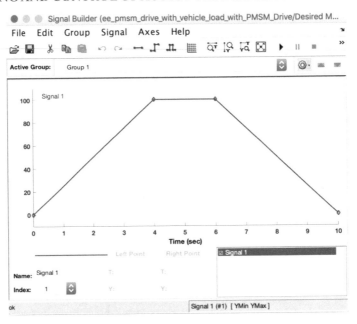

Figure 6.54: Function generator for motor speed.

12. We will now add the controller/motor drive in this model. We have discussed earlier the field-oriented control block that already exists in the Simscape library. Create a subsystem and name it PMSM Controller. Open the subsystem and add the PMSM Field-Oriented Control block from the Simscape library. The block has five inputs and a couple of outputs. As Figure 6.55 shows the five inputs are the reference/desired motor speed in rad/s, the 3-phase measured current in the motor, the measured speed, and the measured angular position of the rotor and the motor voltage setting. Connect three inport blocks as shown in Figure 6.55. Block 1 will bring in measured information from the motor speed and position that is stored in the meas Goto block. Inport 2 will be connected to the ref_motor_rpm and the rpm is converted using a gain block that multiplies rpm with 2*pi/60. Inport 3 will bring in the measured three phase currents that are stored in the i Goto block. Attach the constant 400 for the Motor voltage. After the connections are made the input side looks like it is shown in Figure 6.55.

13. On the output side of the PMSM field-oriented control block the G terminal outputs the control signal that needs to feed into the G terminal of the 3-phase inverter. This carries the switching signals for the inverter. This data is also stored in the G Goto block. The second output terminal is the visualization terminal that outputs data for visualization. The visualization output is demmuxed into three separate outputs, the iabc—3-phase motor currents, the demand speed (in rad/s), and the demand torque of the motor.

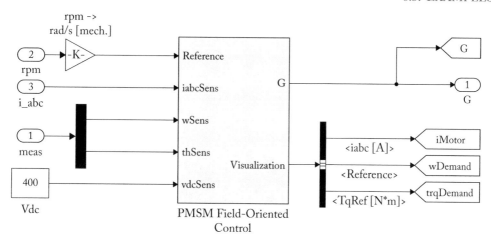

Figure 6.55: PMSM controller subsystem.

14. The input and output terminals of the PMSM controller subsystem are connected to appropriate items as shown in Figure 6.60 using rate transition blocks. Figures 6.56–6.59 show the important settings for this controller.

15. After adding the controller and the signal subsystem (Figure 6.60), the model looks like Figure 6.61.

16. We will now add the final part of the model, data and visualization. Add a new subsystem and name it Signals. Open it and add six From blocks. Name them trqDemand, trqMotor, wDemand, wMotor, G, and iMotor. Use two Mux blocks to couple the two torque signals and the two speed signals. Use a gain block to convert the speed signals from rad/s to rpm and use rate transition blocks for all the signals and connect them to Outport; see Figure 6.62. Connect a scope to the signal subsystem and change the number of channels to 4.

17. The model is now complete and after assembly it should look like Figure 6.63.

18. Set the slope angle to zero on the input side of the Vehicle body subsystem and run the simulation. The results obtained is shown in Figure 6.64.

19. To test the effect of non-zero angles of slopes the slope angle is set to 2° and then to 4° and the simulation is re-run. The results are shown Figures 6.65 and 6.66.

Simulation results from the three plots show some very clear trends. For 0° slope i.e., on level ground as the speed increases the torque demand and the torque output closely matches each other. At the steady-state portion of the travel the torque is equal to zero and when the

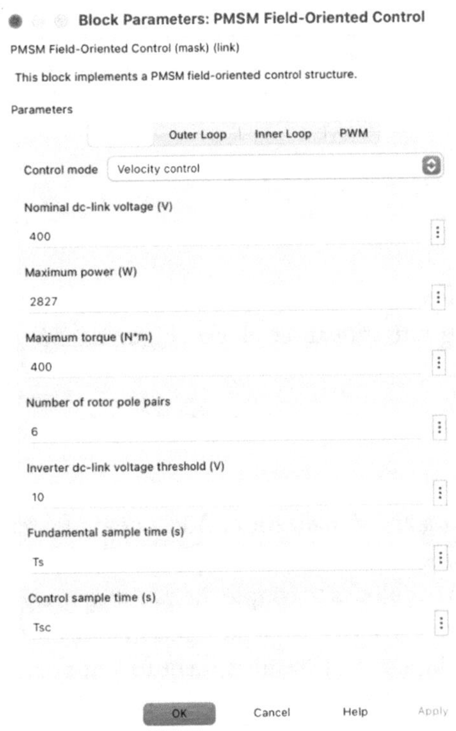

Figure 6.56: PMSM controller setup, general settings.

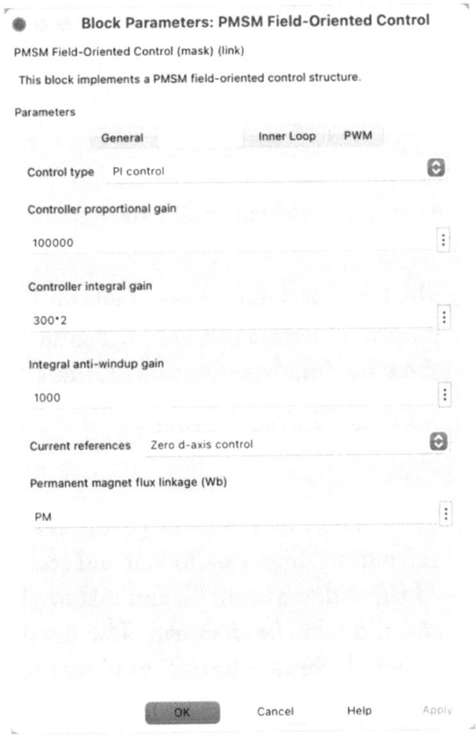

Figure 6.57: PMSM controller setup, outer loop settings.

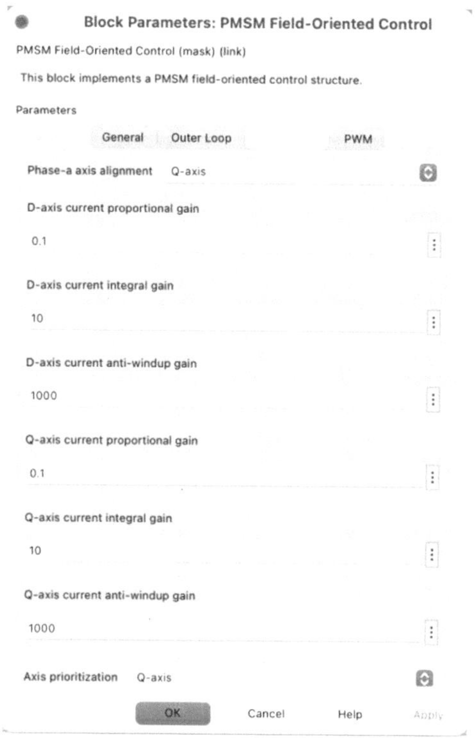

Figure 6.58: **PMSM** controller setup, inner loop settings.

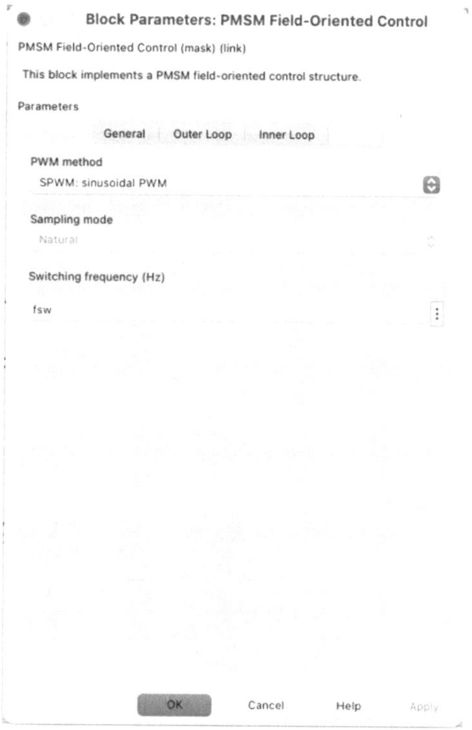

Figure 6.59: **PMSM** controller setup, PWM settings.

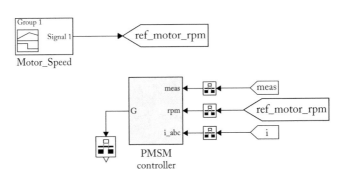

Figure 6.60: PMSM controller and desired motor speed.

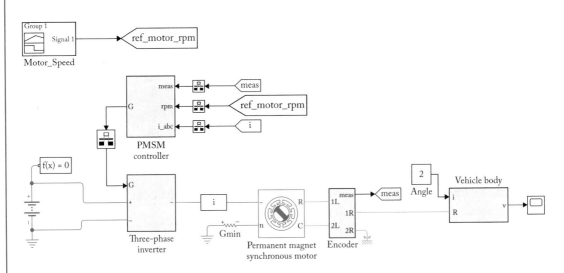

Figure 6.61: Entire model without the output results scope.

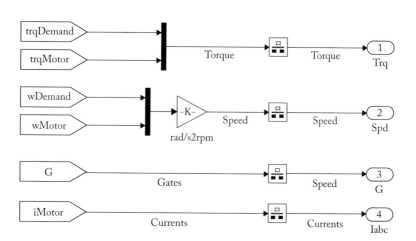

Figure 6.62: **Signals subsystem.**

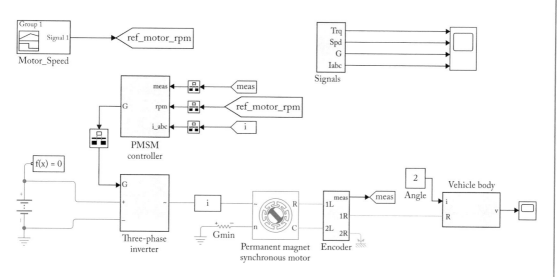

Figure 6.63: **Complete model.**

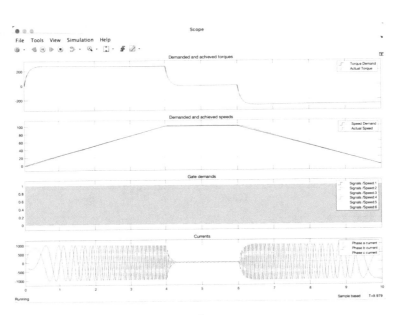

Figure 6.64: Simulation results for 0° slope road.

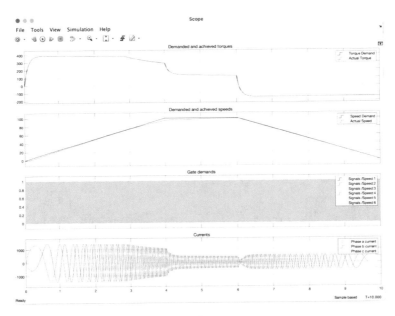

Figure 6.65: Simulation results for 2° slope road.

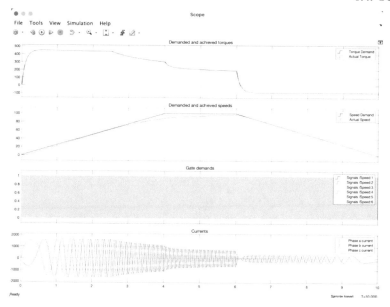

Figure 6.66: Simulation results for 4° slope road.

vehicle is slowing down the torque is negative, i.e., this is when regenerative braking can charge the battery. A similar trend is seen in the other two plots for 2° and 4° slopes, the primary difference being that for these two cases the torque is not zero when the velocity is steady as the vehicle has to overcome significant gravity load. The motor currents in all three cases show appropriate variation as the vehicle transitions from one part of the travel to another. In the zero- and two-degree cases the desired motor speed and the achieved motor speed matches quite well. However, there is some discrepancy in those two quantities in the third case. This could be fixed by tuning the control parameters. That tuning effort has not been attempted here.

6.4   SUMMARY

In this chapter we went through with the modeling of some AC motors that are used as traction machines for HEV and EV applications. The goal was to establish the modeling process for these machines and highlight their typical characteristics. In actual traction applications AC motors are used and they are controlled in a precise manner so that optimum performance can be ensured for all conditions. In the next chapter we will consider the modeling of different vehicle architectures for HEV vehicles.

CHAPTER 7

Modeling Hybrid Vehicle System Architecture

7.1 INTRODUCTION

In earlier chapters in this book, we have discussed different subsystems that make up HEVs and EVs. This chapter is devoted to exploring the entire system model. Here, all parts of the system will come together as one functioning complex system. There are several well-known system architectures that are used in HEV design. In this chapter we will discuss some of them purely from the modeling perspective. In earlier chapters we explored subsystems in great detail to understand and model how certain parts of the overall system work. Here we make some simplifications of those same subsystems to devote more attention to the entire system model. In very broad terms, the hybrid vehicle world is divided into three different architectures: series, parallel, and series/parallel. There are additional architectures or variations which have been named complex hybrid, diesel hybrid, etc. Figure 7.1 shows the general layout and connectivity of components in each of the cases. In this chapter we discuss models of a series architecture, a parallel architecture, and a combination that is aided through a planetary gear train. Before we get into the discussion of the models themselves, we will briefly outline key features of each of the architectures. We are not making any attempt to get into all possible details on these topics since our goal is model development and simulation. The reader is encouraged to refer to other references on this topic for more detailed discussion.

7.2 HEV ARCHITECTURES

It is well known that internal combustion engines are not very efficient at energy conversion. Also, its efficiency varies with speed and torque outputs. The overall efficiency is best at the middle speed and high torque region. This means that ICEs are very inefficient in urban driving conditions in stop and go traffic. On the other hand, electric motors produce high torque in the low-speed operations and their overall efficiency even in low-speed operation is fairly high. If the vehicle is propelled by the electric motor in low-speed operations, its fuel economy can be improved significantly. Also, the presence of the motor can mean that the ICE can be operated at the optimum condition independent of the road load. The motor also allows regenerative braking, when the vehicle's kinetic energy can be retrieved and used to recharge the battery. At

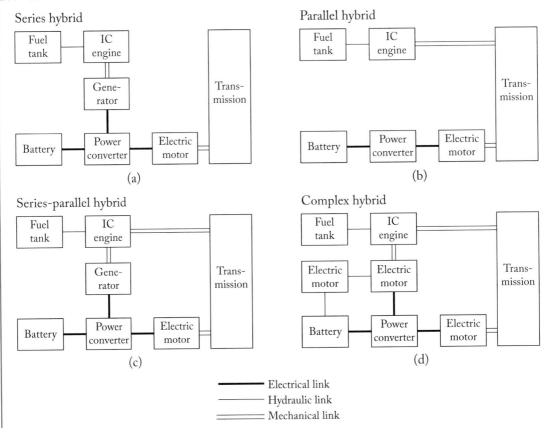

Figure 7.1: Different HEV architectures.

higher driving speeds or on highways the ICEs provide sustainable power and longer driving range.

7.2.1   SERIES ARCHITECTURE

In the case of series architecture, the engine and the battery are two power sources. The power generated from the engine is used to drive a generator which in turn drives a motor as well as charge the battery with the excess power that is generated. Figure 7.2 shows a schematic of the series architecture. The electric motor drives the vehicle using electricity generated by the generator or electricity drawn from the battery. As the engine is decoupled from the wheels the engine speed can be controlled independent of the vehicle speed. This not only simplifies the control but also allows engine operation at optimum efficiency. This also allows a lot of flexibility of locating the engine, since a transmission is not needed in a series HEV. The series hybrid is

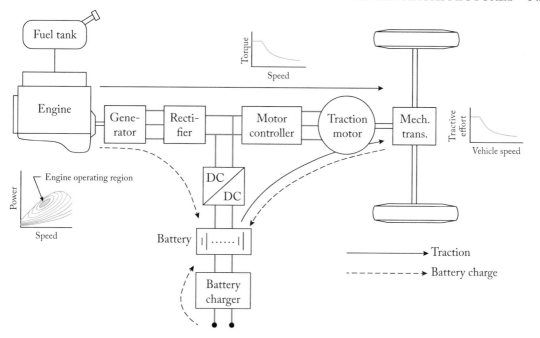

Figure 7.2: Series architecture.

designed to operate in various modes such as: battery alone, combined power, engine alone, power split, stationary charging, and regenerative mode.

7.2.2   PARALLEL ARCHITECTURE

In the case of parallel architecture, the engine and battery can both deliver power to the wheels. The ICE and the motor outputs are mechanically coupled to the final drive by various mechanisms such as a clutch, belts, pulleys, or gears. Figure 7.3 shows a schematic of the parallel architecture. In the parallel architecture the need for a transmission exists. In certain design versions the transmission is used prior to coupling the ICE and the Motor outputs and others it is used after the coupling of the two outputs. The parallel hybrid is designed to operate in various modes such as: motor alone, combined power, engine alone, power split, stationary charging, and regenerative mode.

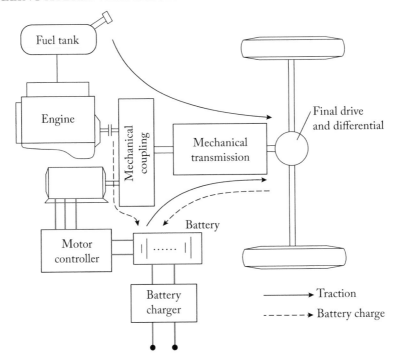

Figure 7.3: Parallel architecture.

7.2.3   SPLIT POWER SERIES/PARALLEL MODE

One of the ways that power is split between the different power sources used in parallel is through the use of a planetary gears. A planetary gear train consists of three gears, sun, ring, and planets, with the planets being mounted on a carrier. Figure 7.4 shows a schematic of a planetary gear train. The planetary gear train has three ports, which means it has three input/output points and the speeds of rotation of the three components, sun, carrier, and the ring gears are related through a kinematic relationship. The kinematic relationship for the planetary gears is given by:

$$\frac{\omega_s - \omega_c}{\omega_r - \omega_c} = -k = -\frac{Nr}{Ns} \tag{7.1}$$

$$\omega_s + k\omega_r - (1+k)\omega_c = 0 \tag{7.2}$$

$$\omega_c = \frac{k}{(1+k)}\omega_r + \frac{1}{(1+k)}\omega_s, \tag{7.3}$$

where the subscripts s, c, and r represent sun, carrier, and the ring, respectively; N is the number of teeth and ω is the angular velocity in radians per second. Similarly, the torque and power equations are given by:

$$T_c = T_r + T_s \tag{7.4}$$

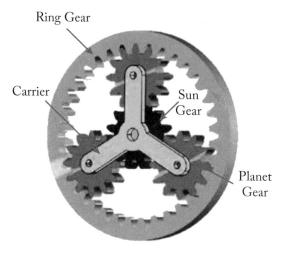

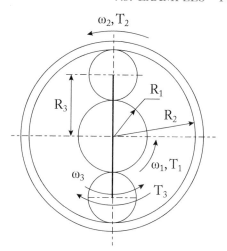

Figure 7.4: Schematic of a planetary gear train.

$$T_c\omega_c = T_r\omega_r + T_s\omega_s. \tag{7.5}$$

In this split power arrangement using a planetary gear train, the engine input is attached to the carrier, the motor is attached to the ring and the generator is attached to the sun. And using appropriate control algorithm the many different operating modes are achieved. Figure 7.5 shows a schematic of some of the operating modes and the relative speed of the three devices, engine, motor, and the generator. The four operating modes shown in the picture are vehicle launch or starting, normal cruise, engine cranking, and power boost/acceleration.

To illustrate the operation of different vehicle architectures we have used three examples in this chapter. The first one shows a series architecture, the second one is a parallel architecture with a motor and an engine driving the vehicle through a single input to the engine. The mode of coupling of the two inputs is not specifically included in this second example. In the third example, a power split device, a planetary gear train, is included to couple the engine, motor, and a generator. In all three examples a standard speed profile is used to show the vehicle operation.

7.3   EXAMPLES

Following are a handful of tips to add new elements into a model.

Tips for adding model elements

1. Use Quick Insert to add the blocks. Click in the diagram and type the name of the block. A list of blocks will appear and you can select the block you want from the

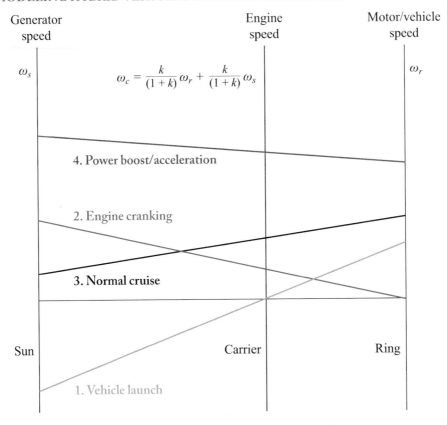

Figure 7.5: Relative speeds of the different components during different operating modes.

list. Alternatively, the Open Simscape Library block can be used to look though the library of all blocks and pick the appropriate one.

2. After the block is entered, a prompt will appear for you to enter the parameter. Enter the variable names as shown below.

3. To rotate a block or flip blocks, right-click on the block and select Flip Block or Rotate block from the Rotate and Flip menu.

4. To show the parameter below the block name, see Set Block Annotation Properties in the Simulink documentation.

7.3.1   EXAMPLE 7.1: MODELING THE SERIES HYBRID ELECTRIC DRIVE

In the series architecture, the engine is driven at a setting where its efficiency is the highest. The engine drives the generator which in turn drives the motor and any excess power is used to recharge the battery. Also, during acceleration, when additional power is needed, the battery supplies this additional power to drive the motor. So, in the model we will be using a battery setup, two motor/generator modules, an engine module and an automobile module with associated road, slope, and wind loads. In the previous chapter we used some of these components/subsystems and we will duplicate the same components for our use here.

Simscape offers a variety of motor model blocks that can be used to model many types of DC and AC Motors. In Chapter 6 we used two different PMSM motor models. Here we will use the Simplified PMSM drive block which was used in Chapter 6. As its name suggests, it is a simplified block that is good for system level modeling. It uses a DC input and is designed to provide torque and speed outputs that are typical of a PMSM motor. The block uses a steady-state torque speed relationship curve to determine the motor output (instead of the fundamental electromagnetic equations). So, implementing a system model using this motor is easier, hence the choice. Figure 7.6 shows this motor model location in the Simscape library. The torque speed characteristic of this motor was discussed in Chapter 6.

Steps:

1. Type **ssc_new** in the Matlab command window to open a new model file.

2. Copy and paste the battery subsystem that was used in Chapter 2 and in Example 6.3.1. The battery subsystem is shown in Figure 7.7.

3. The battery subsystem has a generic battery model from the library. Its settings are kept the same as was in Example 6.3.1. It also has a calculations subsystem shown in Figure 7.8.

4. Calculations subsystem stores the calculations in Goto blocks called Pe_battery, Chg_battery, and Ploss_battery, respectively.

5. Add a DC-DC convertor block to the model. This block was also used in Example 6.3.1. So, copy and paste the same block here. This block models a buck-boost convertor without modeling the actual circuit. The settings used for the DC-DC convertor are kept exactly the same as they were in Example 6.3.1.

6. For this example, we will use one of the standard drive cycles used in vehicle testing. It is called the FUDS. This is the EPA Federal Urban Test procedure and the vehicle speed data is available easily in open source. This data is shown in Figure 7.9 and is available in the form of a spreadsheet. The speed data is in miles per hour. So, we will need to ensure it is converted into m/s. Add a From Spreadsheet block and link the spreadsheet file to it as shown in the block options in Figure 7.10.

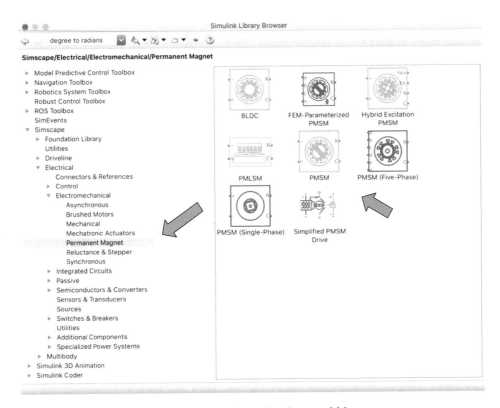

Figure 7.6: Locating the PMSM motor block in the electrical library.

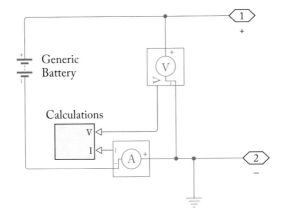

Figure 7.7: Building the battery model.

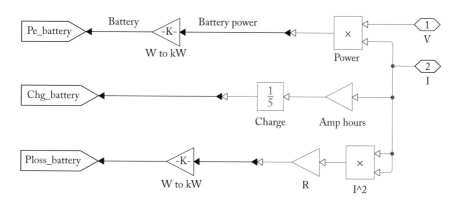

Figure 7.8: Subsystem calculations.

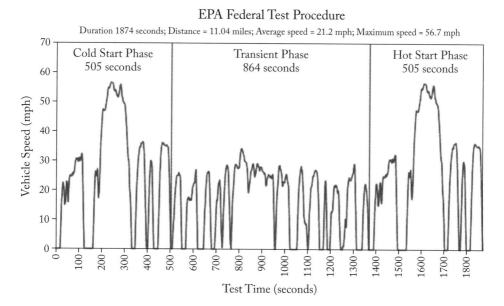

Figure 7.9: Vehicle speed cycle.

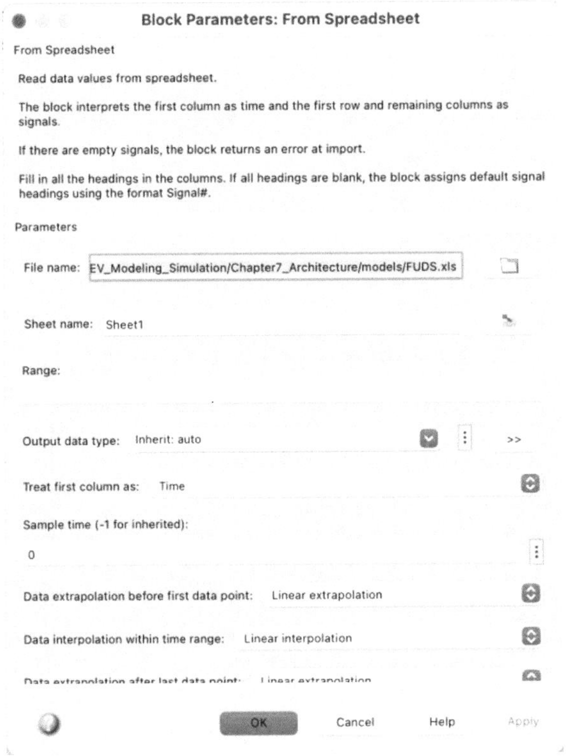

Block Parameters: From Spreadsheet

From Spreadsheet

Read data values from spreadsheet.

The block interprets the first column as time and the first row and remaining columns as signals.

If there are empty signals, the block returns an error at import.

Fill in all the headings in the columns. If all headings are blank, the block assigns default signal headings using the format Signal#.

Parameters

File name: EV_Modeling_Simulation/Chapter7_Architecture/models/FUDS.xls

Sheet name: Sheet1

Range:

Output data type: Inherit: auto   >>

Treat first column as: Time

Sample time (-1 for inherited):

0

Data extrapolation before first data point: Linear extrapolation

Data interpolation within time range: Linear interpolation

Data extrapolation after last data point: Linear extrapolation

OK   Cancel   Help   Apply

Figure 7.10: From spreadsheet block settings.

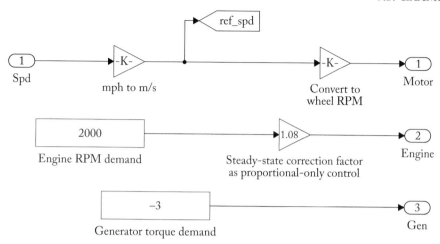

Generator torque demand has been set so that at the starting condition
the combustion engine is providing just enough power to keep the vehicle
moving at constant speed, and hence battery power is close to zero. The
engine RPM demand is set to 2000 rpm for good efficiency.

Figure 7.11: **Strategy subsystem.**

7. Now add a Subsystem and call it Strategy. Earlier, we talked about the strategy of a series
   hybrid. The engine is run at an optimum efficiency setting at a constant speed. In this
   example, the engine speed is set at 2000 rpm. Similarly, the torque demand of the generator
   is also set in a way that at starting all the power is coming from the engine and no battery
   power is expended. This subsystem will handle three different settings; the vehicle speed
   demand, the engine setting, and the torque demand from the generator. Figure 7.11 shows
   the strategy subsystem contents.

8. Connect the speed input from the spreadsheet block to the Strategy subsystem and
   the three outputs from the Strategy block to three Goto blocks named ref_motor_rpm,
   ref_engine_speed, and ref_generator_torque, respectively. To convert the vehicle speed to
   motor rpm we use two gain blocks in the Strategy subsystem. The first one converts mph to
   m/s by multiplying the mph with 0.44074 and then it is converted to motor rpm by multiplying with (60/(2pi)/0.3) where 0.3 m is the wheel radius. In this example the generator
   torque demand is held at −3 Nm.

9. The motor used in this model will be identical to the motor subsystem used in Example 6.3.1. Figure 7.12 shows the motor subsystem and Figures 7.13–7.16 show different
   parts of this subsystem. Figures 7.17–7.19 show the different motor parameter settings
   used in this example.

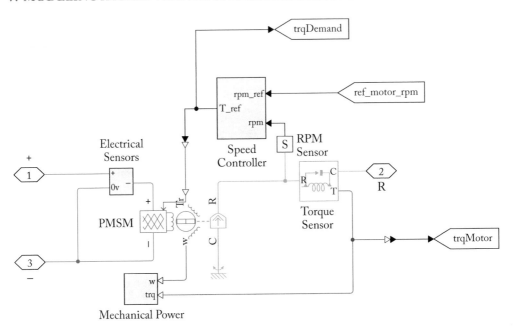

Figure 7.12: Motor subsystem.

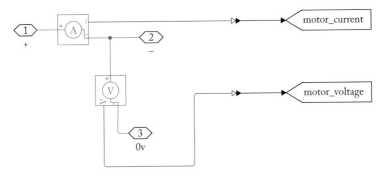

Figure 7.13: Subsystem electrical sensors.

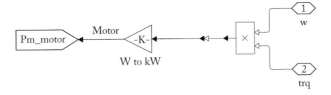

Figure 7.14: Mechanical power subsystem.

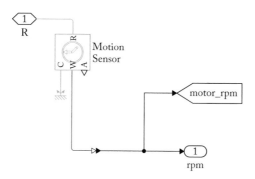

Figure 7.15: **RPM sensor subsystem.**

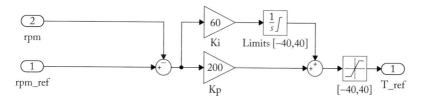

Figure 7.16: **Speed controller subsystem.**

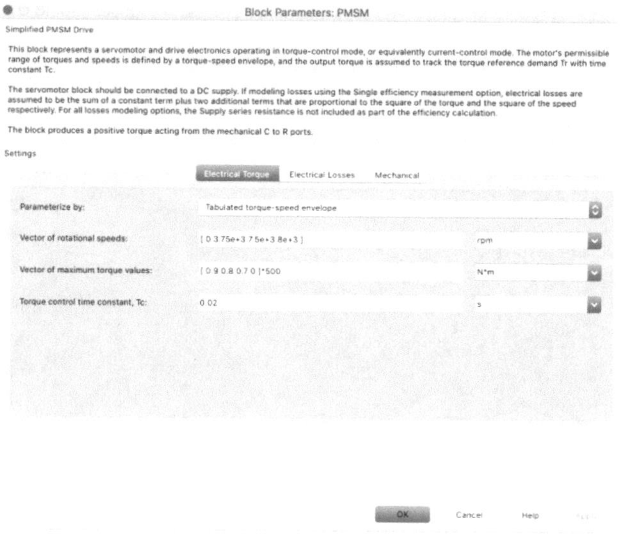

Figure 7.17: **Simplified PMSM drive, electrical torque.**

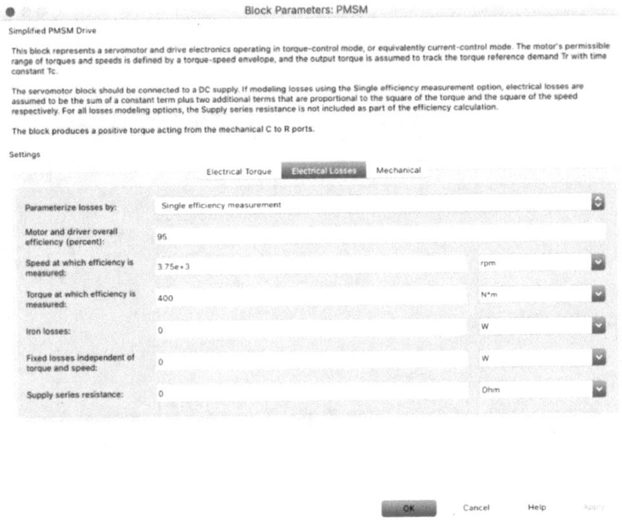

Figure 7.18: Simplified PMSM drive, electrical losses.

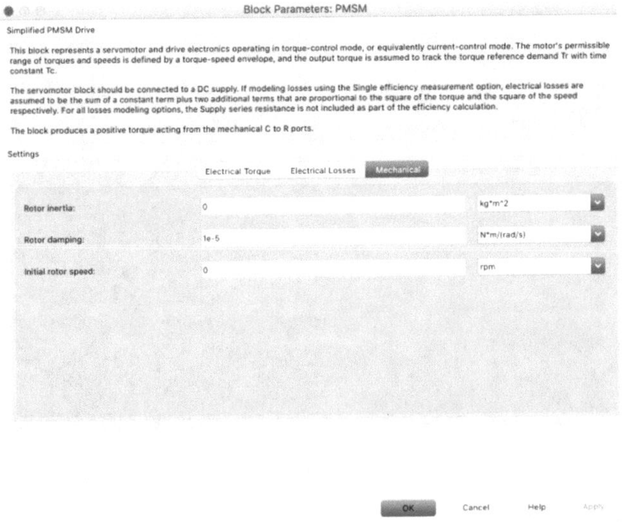

Figure 7.19: Simplified PMSM drive, mechanical.

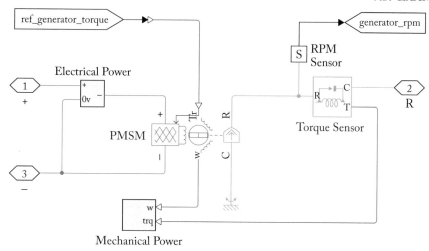

Figure 7.20: Generator subsystem.

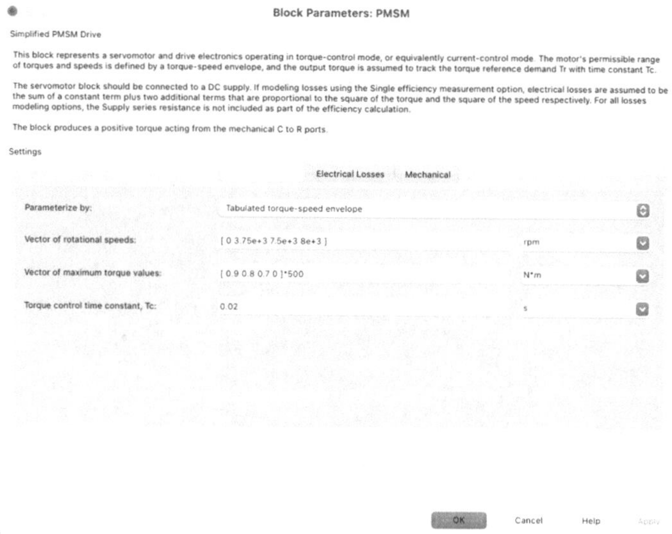

Figure 7.21: Generator setting for electrical torque.

10. The PMSM motor that is used in the motor subsystem is also the same machine that is used in the generator subsystem. The generator subsystem is shown in Figure 7.20. And the different settings for the generator are shown in Figures 7.21–7.23. The rest of this subsystem is built in a manner similar to the motor subsystem. The demand torque is determined by the input demand specified earlier. The speed and torque as well as the

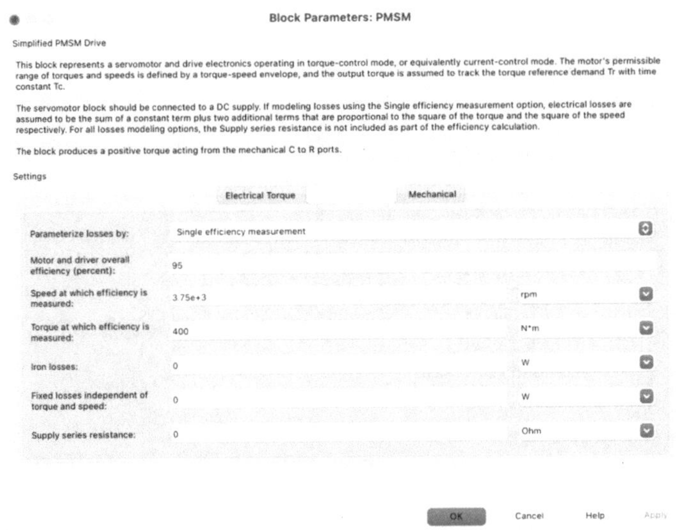

Figure 7.22: Generator settings for electrical losses.

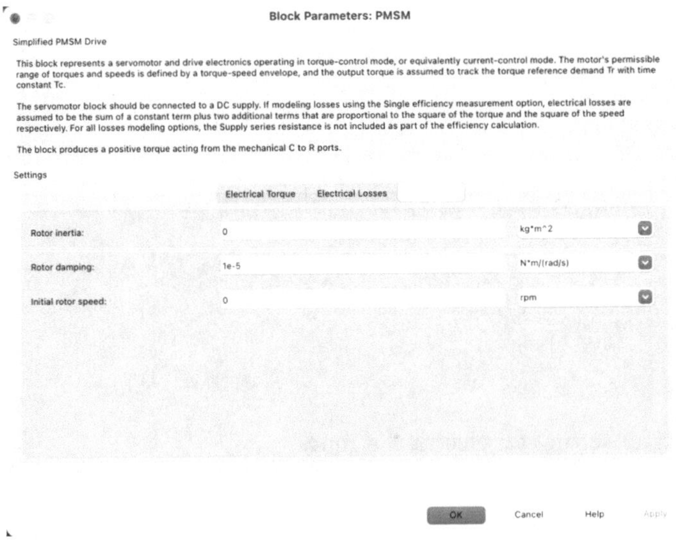

Figure 7.23: Generator settings for mechanical components.

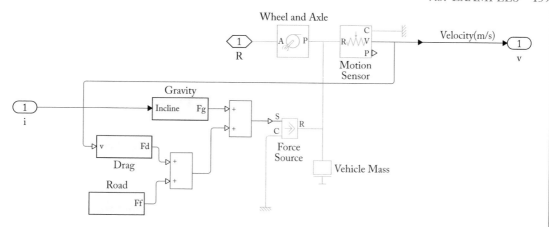

Figure 7.24: Vehicle body subsystem.

voltage and current are used in the same way as in the case of the motor to compute electrical and mechanical power. And they are saved in Goto blocks named Pm_generator for the mechanical power and Pe_generator for the electrical power.

11. We will now add the load/resistance force for the vehicle, i.e., the vehicle model. The development of the vehicle model comprising of the different load types was described in Example 6.3.1. The same exact vehicle model is used here so we will not be describing its details again. It was called the Vehicle Body subsystem in that previous example. We copy and use it here. All the parameters used in that subsystem remain unchanged here as well.

12. Figure 7.24 shows the entire Vehicle Body subsystem.

13. After assembling everything we have created thus far, the model looks like the one shown in Figure 7.25.

14. We will now need to create the sole new subsystem for this model, the engine. The engine subsystem is built around the generic engine block that is available in the Simscape library. Add a new subsystem in the model and call it Engine. Open the subsystem and add the engine Block from the library. The engine block has two terminals on the left and three on the right. The T terminal receives Throttle position information which is a number between 0 and 1 indicating fraction of throttle opening. The B terminal is connected to mechanical ground or base. On the right side the P terminal provides Power information for the engine, the FC terminal provides fuel consumption rate information and the F terminal is the actual mechanical output of the engine. Figure 7.26 shows the engine subsystem with the engine block along with other additional blocks. The output from the power terminal is stored in a Goto block named Pm_engine. The output of the

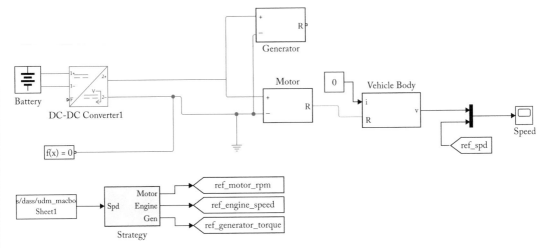

Figure 7.25: Assembled model.

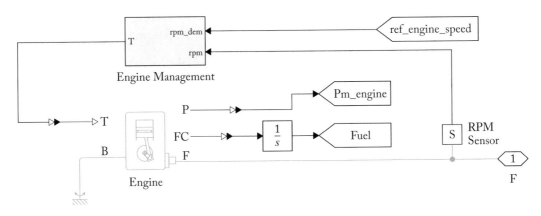

Figure 7.26: Engine subsystem.

FC terminal is integrated and then stored in a Goto block named Fuel Consumption to track the total fuel consumption over time.

15. The engine block allows the user to set parameters for engine torque, dynamics, fuel consumption, limits, and speed control. Figures 7.27–7.31 show the details of the settings used in this case.

16. The speed of rotation is measured at the engine output using the rpm sensor subsystem and this is fed into the engine management subsystem that provides the throttle signal to the engine. The rpm sensor subsystem is shown in Figure 7.32.

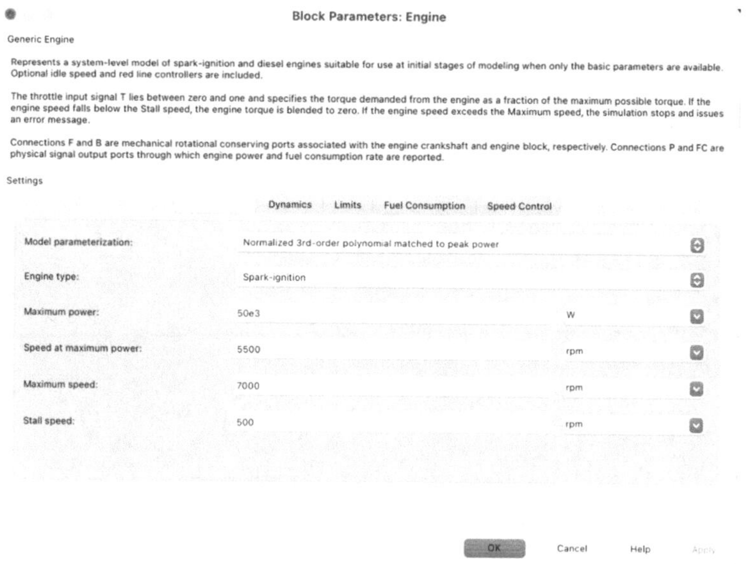

Figure 7.27: Engine torque setting.

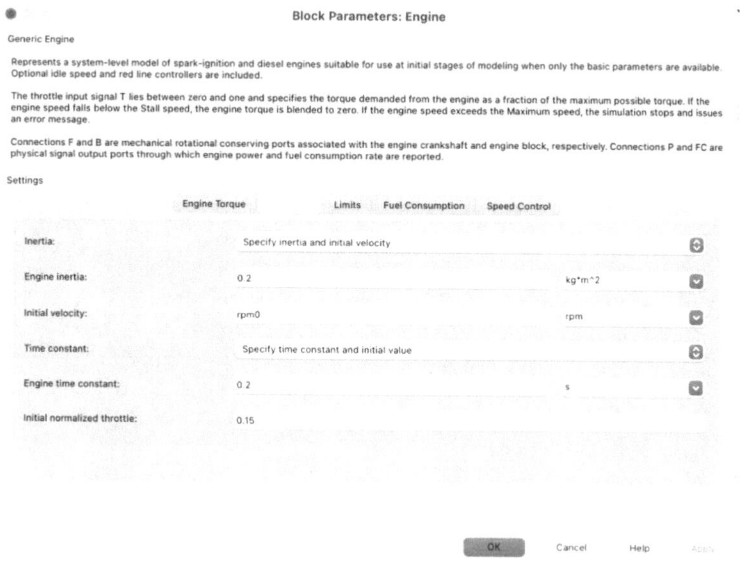

Figure 7.28: Engine dynamics setting.

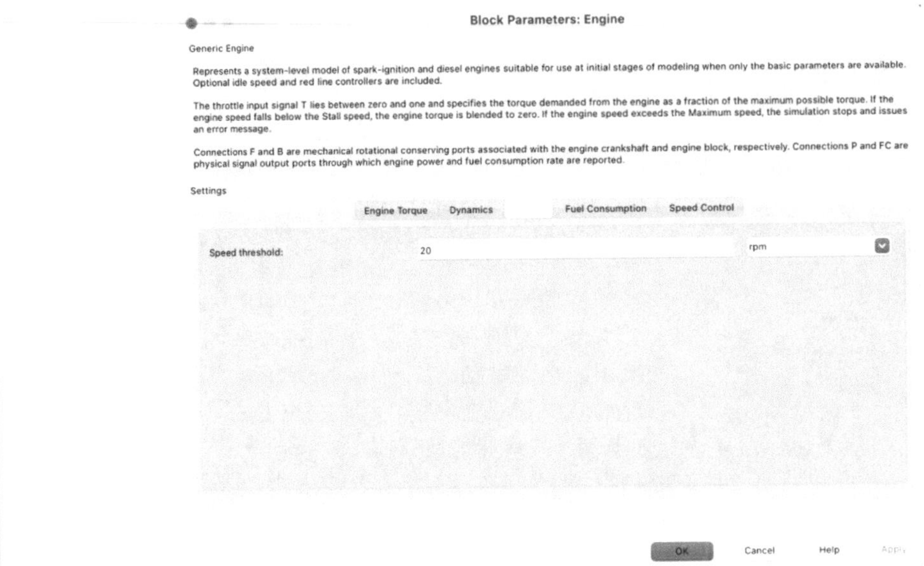

Figure 7.29: Engine limits setting.

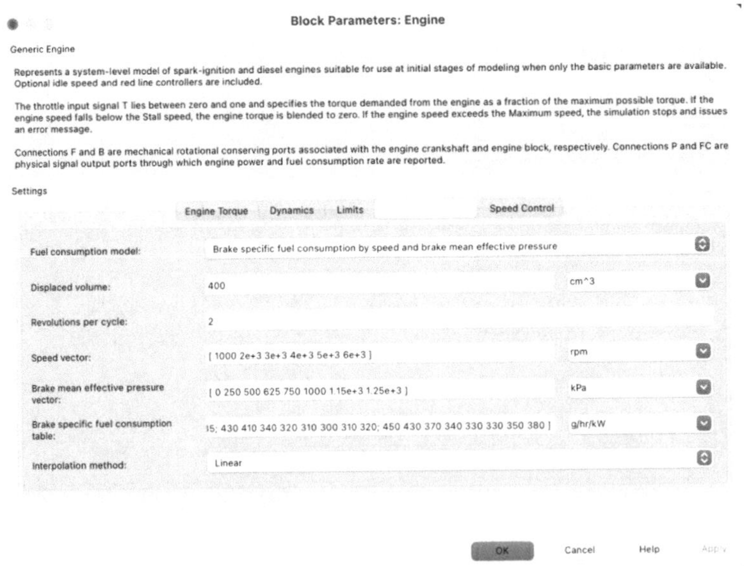

Figure 7.30: Engine fuel consumption settings.

Figure 7.31: Engine speed control setting.

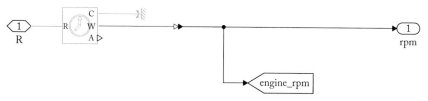

Figure 7.32: Engine speed sensor subsystem.

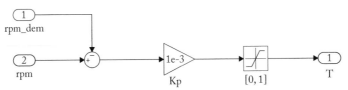

Figure 7.33: Speed error to compute throttle position.

17. Figure 7.33 shows the engine management subsystem which receives the engine rpm data as well as the rpm demand data (from speed input). The difference of the two is rpm error or difference. This is multiplied by a proportionality control constant and is bounded by a lower limit of 0 and an upper limit of 1. The result is the throttle position signal, a number between 0 and 1.

18. Add one more subsystem to the model and name it Scope. This will contain most of the output plotting information. Figure 7.34 shows the content of this subsystem. Data from the entire model was stored in many Goto blocks. The corresponding From blocks are used to retrieve the data so that they can be displayed in graphical form on scopes.

19. Assemble the model so that the final model resembles Figure 7.35.

20. The model is simulated for two situations, driving on a flat road, i.e., a road with zero slope and driving on a road with a slope of 2°.

21. The results from these two sets of simulation are shown in Figures 7.36–7.47.

When the results from the 0° and 2° slope options are compared a few items remain unchanged. For example, the desired and actual velocity plots, since a velocity requirement was prescribed, remain unchanged. The engine is set to run at a constant speed and the generator demand was the same in both cases so the engine and generator behavior remains unchanged. In both cases, the motor is responsible for all the variations in speed and that is reflected in the motor behavior. All the extra energy comes from the battery. So, the two battery plots (state of charge and battery power losses) show distinct differences between the two situations, naturally more power/energy is spent when the vehicle is driving up a slope vs. when it is on flat ground. And the fuel consumption in the engine in both cases remain the same since the engine is running at the same setting for the two situations considered here.

7.3.2   EXAMPLE 7.2: MODELING THE PARALLEL HYBRID ELECTRIC DRIVE

In the parallel hybrid architecture, the engine and the electric motor work as two separate sources of power and they are both connected to the drive shaft. Based on the power demand from the system, the power coming from the different sources are divided up accordingly. In the model

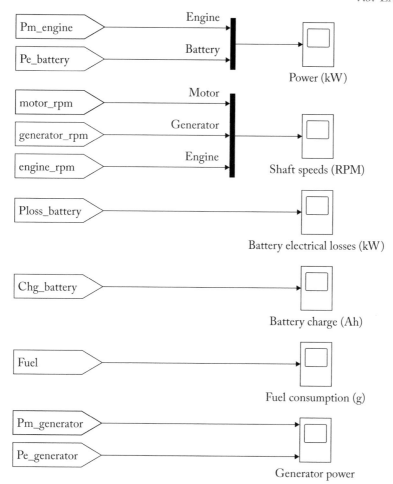

Figure 7.34: Scopes subsystem.

presented here we will use some of the same subsystems used in the series hybrid model, such as the battery and DC-DC converter, the motor, the vehicle model, as well as the engine. The engine will be connected to the drive shaft through a gearbox. The motor will be connected to the same drive shaft as well. To keep things simple we are connecting the two sources without any other additional mechanism. This will mean that the motor speed and the speed of the gearbox outlet will be the same. And the drive cycle used in this simulation is the same city drive cycle used in Example 7.3.1. The only critical difference is how the subsystems are connected and the absence of the generator.

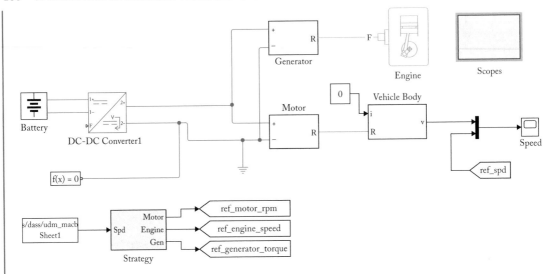

Figure 7.35: Final model.

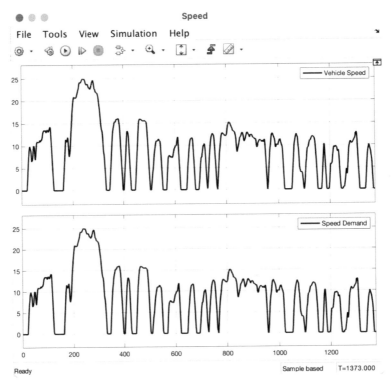

Figure 7.36: Desired and actual vehicle speed at 0° slope.

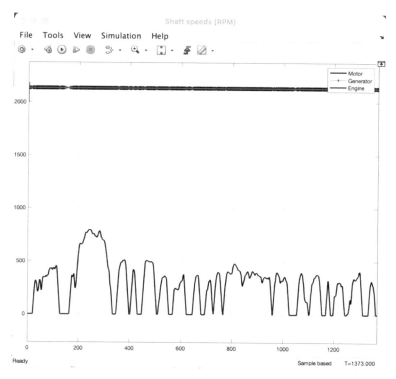

Figure 7.37: Speed comparison of the three sources at 0° slope.

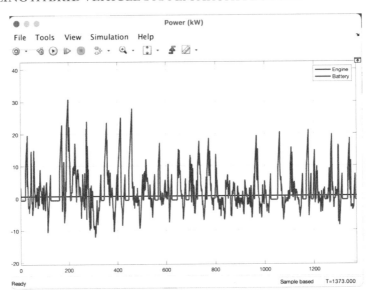

Figure 7.38: Engine and battery power at 0° slope.

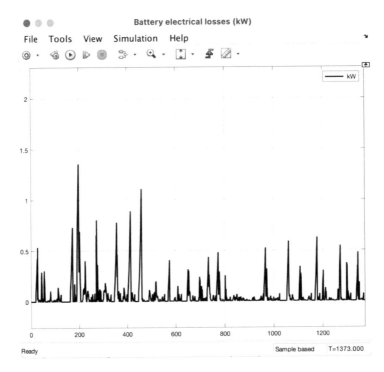

Figure 7.39: Battery losses during a 0° slope.

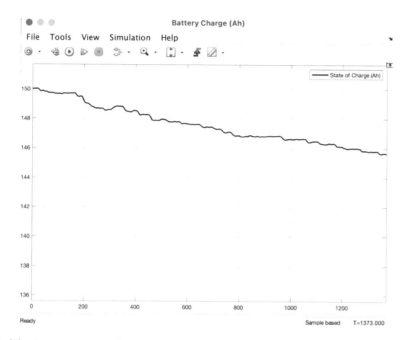

Figure 7.40: The battery state of charge in the model for a 0° slope.

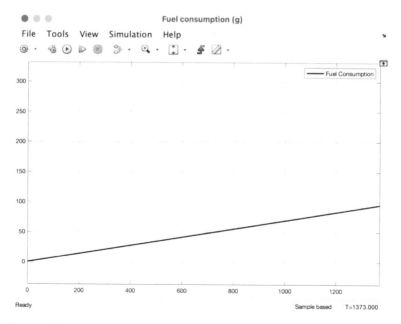

Figure 7.41: Engine fuel consumption at 0° slope.

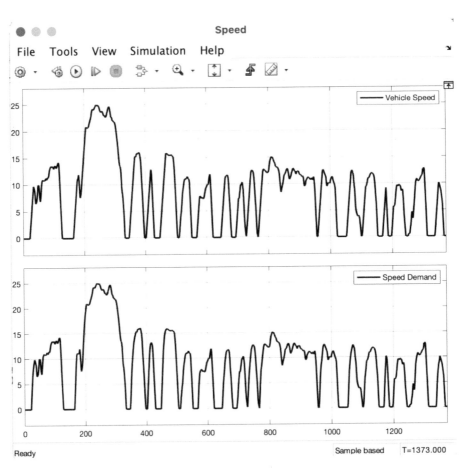

Figure 7.42: Desired and actual vehicle speed at 2° slope.

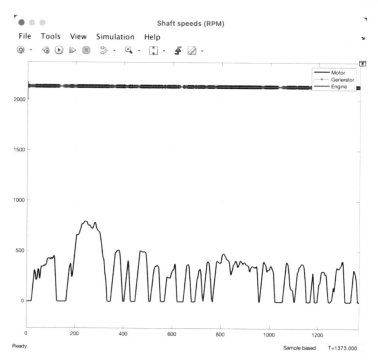

Figure 7.43: Speed comparison of the three sources at 2° slope.

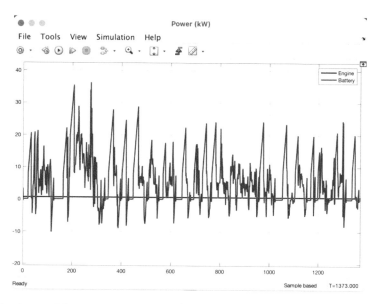

Figure 7.44: Engine and battery power at 2° slope.

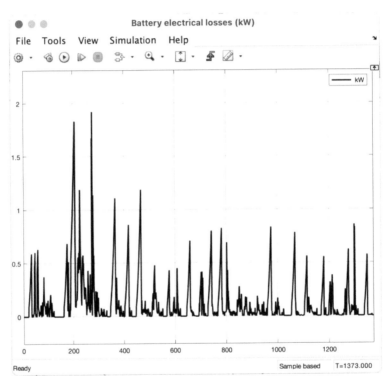

Figure 7.45: Battery losses during a 2° slope.

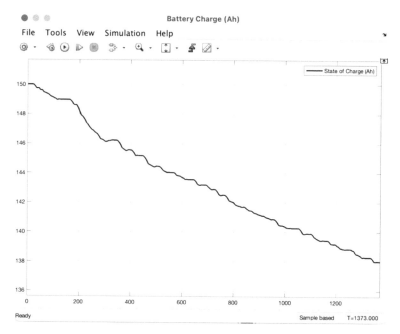

Figure 7.46: The battery state of charge in the model for a 2° slope.

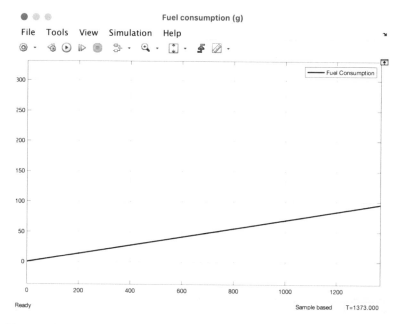

Figure 7.47: Engine fuel consumption at 2° slope.

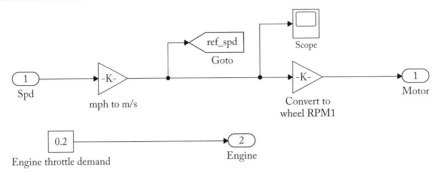

Figure 7.48: Strategy for vehicle speed and engine operation.

Steps:

1. Type **ssc_new** in the Matlab command window to open a new model file.

2. Copy the battery subsystem that was used in Example 7.3.1. Add the DC-DC convertor block from Example 7.3.1.

3. For the standard drive cycle add the From Spreadsheet block from Example 7.3.1.

4. Also add the Strategy subsystem block from Example 7.3.1. Open the Strategy subsystem, keep the motor strategy identical but remove the strategy for the generator since we will not have a generator in this system. Also, modify the strategy for the engine. In Example 7.3.1 the engine was set to run at a constant speed. The way the two parallel paths are connected will not allow for the engine to run at a constant speed. Instead, we will specify the throttle position of the engine to a value between 0 and 1. We choose 0.2 and include it in the strategy. This value is placed in a Goto block called ref_throttle to be used in the engine subsystem. The engine strategy details is shown in Figure 7.48 and the strategy subsystem is shown in Figure 7.49.

5. Now add the Motor Subsystem, the Vehicle subsystem and the scopes. These are identical to the ones in Example 7.3.1. Assemble all the systems as shown in Figure 7.50.

6. Add the engine subsystem from the previous model and a gearbox from the library. Open the gearbox settings and set the teeth ratio to 4 and the direction of rotation to be the same as the engine.

7. Open the Engine subsystem and alter the throttle input by connecting it to a From block called ref_throttle and use a terminator block on the output of the rpm sensor as shown in Figure 7.51.

8. Figure 7.52 shows the entire model after all the assembly is completed.

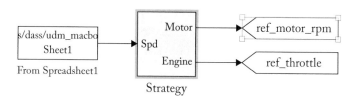

Figure 7.49: The strategy subsystem.

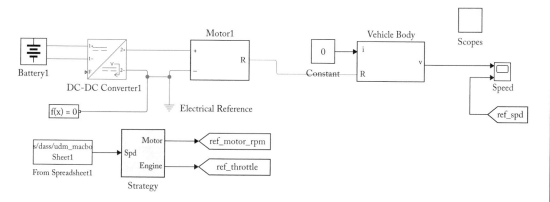

Figure 7.50: Assembled system without the engine.

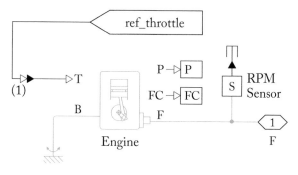

Figure 7.51: Engine subsystem for the parallel architecture.

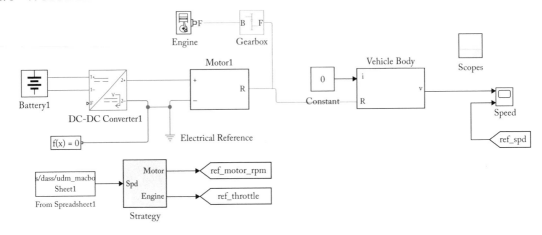

Figure 7.52: The parallel hybrid electric model.

9. Like in the previous example, the model is simulated for two situations, driving on a flat road, i.e., a road with zero slope and driving on a road with a slope of 2°.

10. The results from these two sets of simulation are shown in Figures 7.53–7.64.

When the results from the 0° and 2° slope options are compared a few items remain unchanged. For example, the velocity plots, since a velocity requirement was prescribed, remain unchanged. The engine is set at a constant throttle so the output power from the engine is variable. It changes with the speed change of the vehicle. The speed comparison of the engine output and the motor are also identical in both cases of the slope since this is dependent on the speed demand. The power plot comparison shows that the power output from the engine is pretty identical in both cases but the power output from the motor is different, with more power drawn from the motor/battery in the 2° slope case. This is also reflected in the battery state of charge in both cases. The state of charge reduction is higher in the case of the 2° slope. The fuel consumption in the engine is pretty similar in the two cases.

7.3.3   EXAMPLE 7.2B: PARALLEL ARCHITECTURE WITH DIFFERENT VELOCITY INPUT

Using the parallel architecture arrangement as described in Example 7.3.2 we explore the behavior of the system for a different velocity input. Instead of a standard drive cycle we use a velocity input where the vehicle accelerates steadily from an initial speed of 16–25 m/s. Once the speed reaches this level it stays at that speed for 2 s and then steadily decelerates to 16 m/s. Using this velocity profile we particularly try to see how the energy recovery/regeneration works during the braking or deceleration period as well as the power sharing between the two sources.

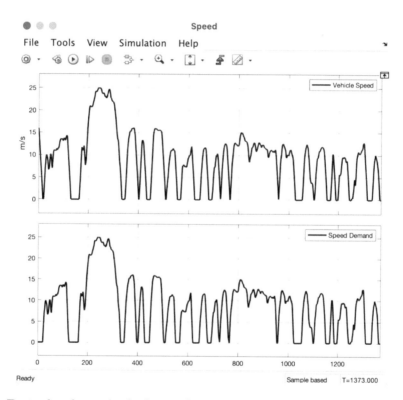

Figure 7.53: Desired and actual vehicle speed at 0° slope.

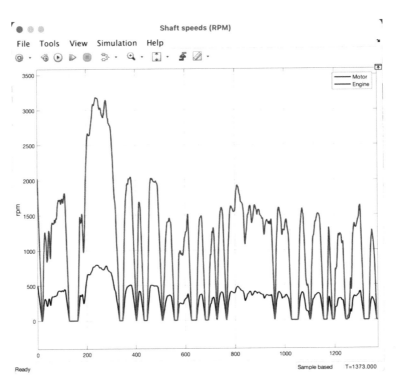

Figure 7.54: Speed comparison of the two sources at 0° slope.

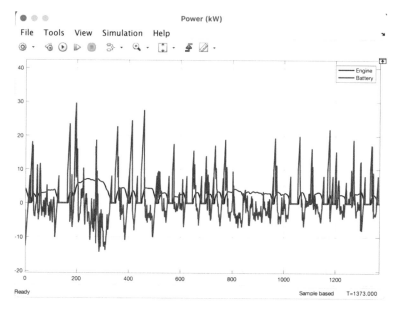

Figure 7.55: Engine and battery power expended at 0° slope.

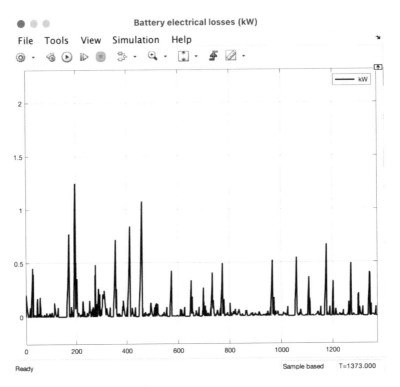

Figure 7.56: Battery losses during a 0° slope.

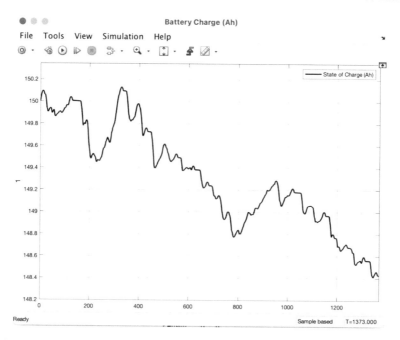

Figure 7.57: The battery state of charge in the model for a 0° slope.

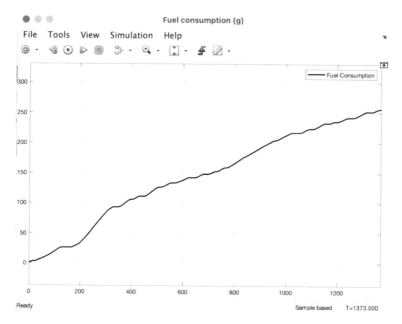

Figure 7.58: Engine fuel consumption at 0° slope.

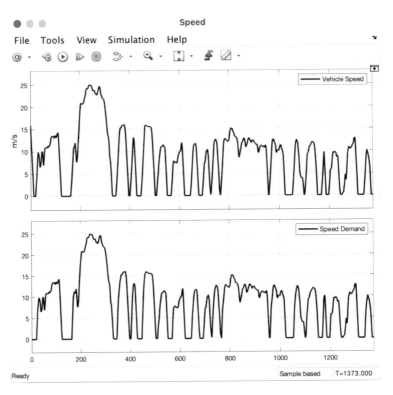

Figure 7.59: Desired and actual vehicle speed at 2° slope.

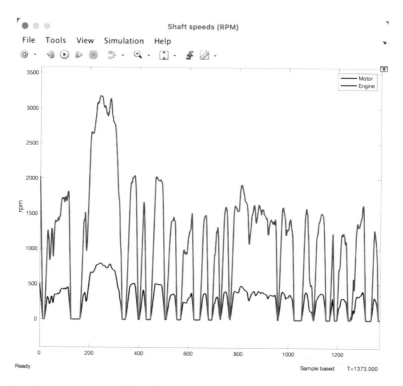

Figure 7.60: Speed comparison of the three sources at 2° slope.

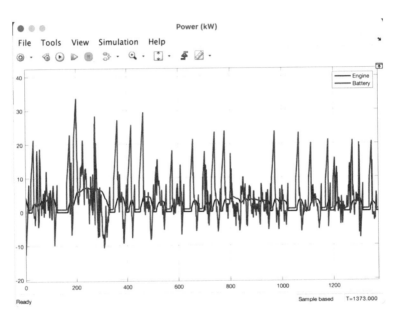

Figure 7.61: Power comparison at 2° slope.

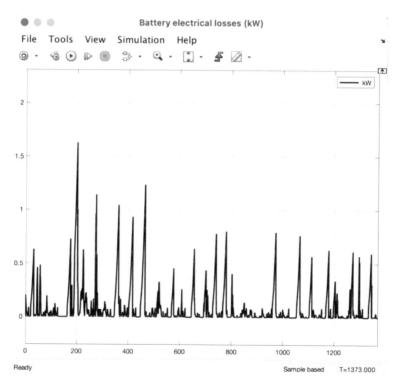

Figure 7.62: Battery losses during a 2° slope.

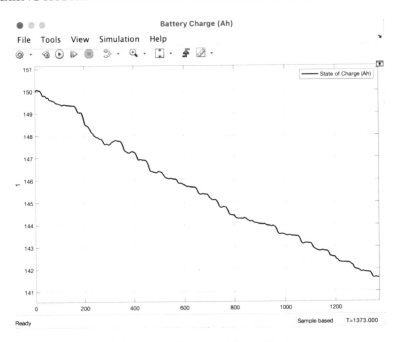

Figure 7.63: The battery state of charge in the model for a 2° slope.

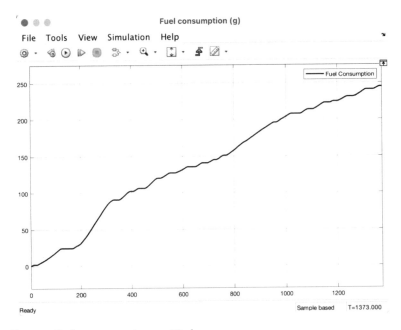

Figure 7.64: Engine fuel consumption at 2° slope.

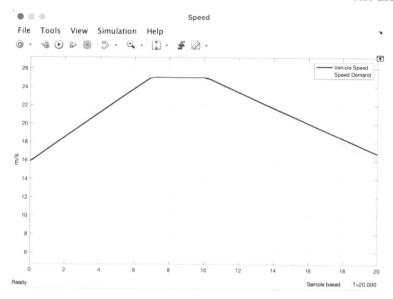

Figure 7.65: Desired and actual speed.

The model and its settings remain the same as was described in Example 7.3.2. Only the throttle setting for the engine was changed from 0.2–0.4 to meet the higher demands from the engine and the analysis is carried out only for 20 s. We are only sharing the results of the simulation here. Figures 7.65–7.70 show different outputs of this analysis. The desired speed and the actual speed show similar behavior though not perfect match. The engine and the motor rpm rise and fall match the pattern of the desired speed. The battery power shows positive and negative portions which indicate the presence of energy recovery from braking during the deceleration phase. The state of charge clearly shows a decrease and subsequent increase as the vehicle goes from accelerating to decelerating phase.

7.3.4 EXAMPLE 7.3: SPLIT POWER PARALLEL MODE HEV MODEL

In the parallel hybrid architecture, the engine input and motor input were both used through the same drive shaft to power the vehicle. A gearbox was used but it was used on the engine side of the parallel path before the output is connected to the common output shaft. No generator is used in the simple parallel architecture. In this example we have used a planetary gear set to manage the power split between three entities, the engine, the motor and the generator. Earlier in this chapter we discussed the power splitting arrangement. That same arrangement is implemented here.

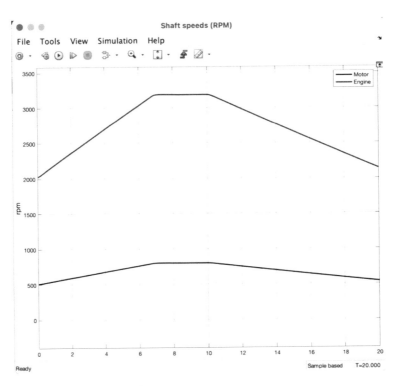

Figure 7.66: Motor and engine speeds.

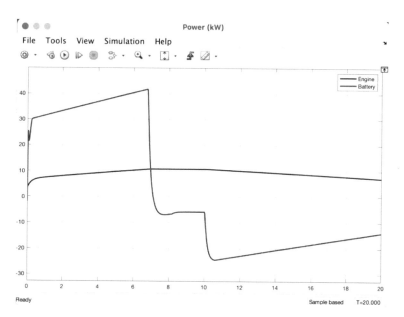

Figure 7.67: Power drawn from the motor and engine.

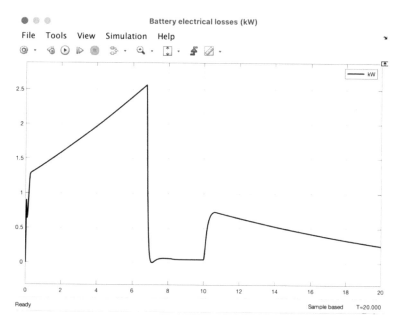

Figure 7.68: Battery energy losses.

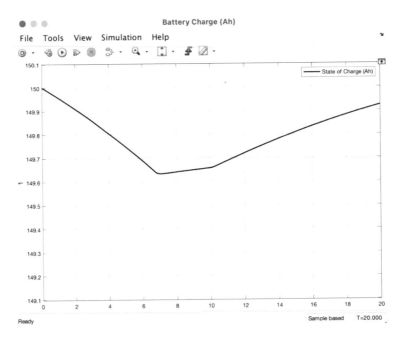

Figure 7.69: State of charge in the battery.

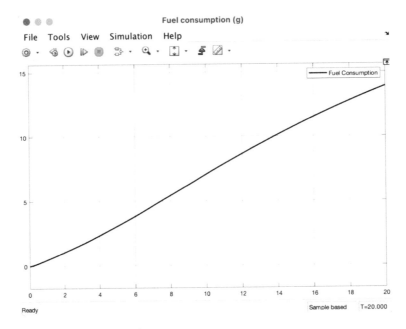

Figure 7.70: Fuel consumption by the engine.

Steps:

1. Type **ssc_new** in the Matlab command window to open a new model file.

2. Copy the battery subsystem and the DC-DC converter block that was used in Example 7.3.1.

3. Like in Examples 7.3.1 and 7.3.2 we will use the standard drive cycle used in vehicle testing. Add the From Spreadsheet block from Example 7.3.1.

4. Also add the Strategy subsystem block from Example 7.3.1. The strategies for the engine, motor, and generator will remain the same in this case.

5. Now add the Motor Subsystem, generator subsystem, Vehicle subsystem, and scopes. These are identical to the previous example as well.

6. Add the engine subsystem from Example 7.3.1.

7. Connect all these systems as shown in Figure 7.71. All the subsystems used so far are the same as in Example 7.3.1.

8. We will now add the planetary gear train that will be used to split the power. Add a new subsystem and name that CRS Planetary (CRS stands for carrier, sun, and ring). Open the subsystem and add a Ring-Planet block and a Sun-Planet block from the library. These two blocks connected together will make a planetary gear. The planetary gear train has three ports which can be used for two inputs, one output, or one input, two outputs.

9. Connect the two gears as shown in Figure 7.72. And set the gear ratios as shown in Figures 7.73 and 7.74.

10. This new subsystem will have three connectors, C, R, and S. Connect the Sun (S) to the generator, the Ring (R) to the motor and the inlet to the vehicle, and connect the engine to the carrier (C). Once assembled the whole system will look like the model shown in Figure 7.75.

11. Like in the previous examples in this chapter, the model is simulated for two situations, driving on a flat road, i.e., a road with zero slope and driving on a road with a slope of $2°$.

12. The results from these two sets of simulation are shown in Figures 7.76–7.85. The desired and actual vehicle speed plots are not included here anymore since there continues to be very good match between those two quantities.

When the results from the $0°$ and $2°$ slope options are compared a few items remain unchanged. For example, the velocity plots remain unchanged, since a velocity requirement of the engine was prescribed, and the velocity demand for the vehicle remain unchanged between

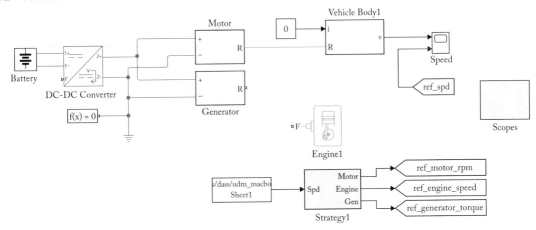

Figure 7.71: Assembled model without all the subsystems that are identical to Example 7.3.1.

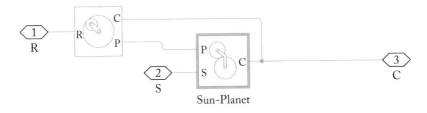

Figure 7.72: Planetary gear subsystem.

the two conditions. The speed comparison of the engine output and the motor are also identical in both cases of the slope since this is dependent on the speed demand and the planetary power split setup. The power plot comparison shows that the power output from the engine is pretty identical in both cases (by design) but the power output from the motor is different, with more power drawn from the motor/battery in the 2° slope case. This is also reflected in the battery state of charge in the two cases. The state of charge reduction is higher in the case of the 2° slope. The fuel consumption in the engine remains similar in the two cases since the engine runs at a constant setting.

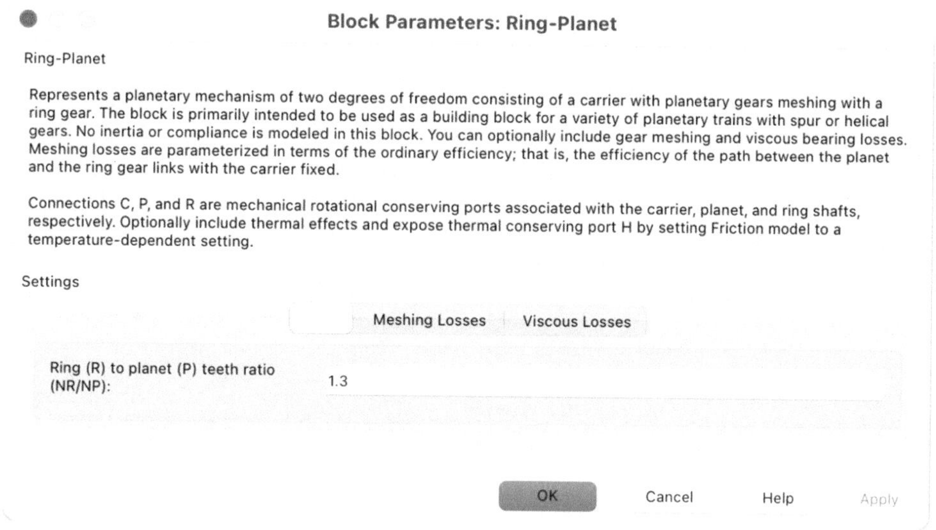

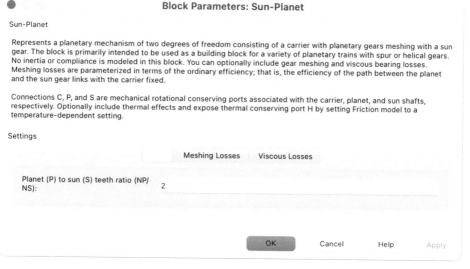

Figure 7.73: Ring-planet gear setting.

Figure 7.74: Sun-planet gear setting.

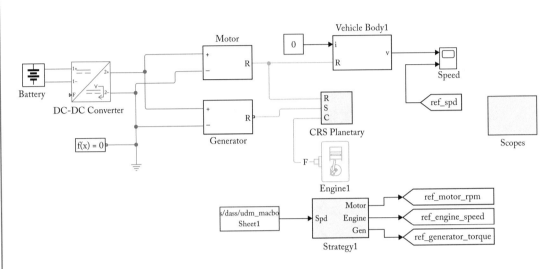

Figure 7.75: The power splitting parallel HEV model.

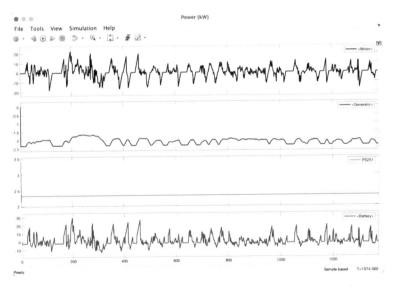

Figure 7.76: Engine, battery, generator, and motor power expended at 0° slope.

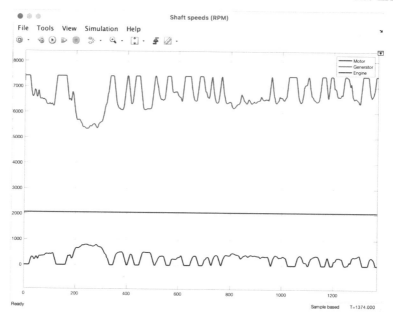

Figure 7.77: Motor, generator, and engine speeds at 0° slope.

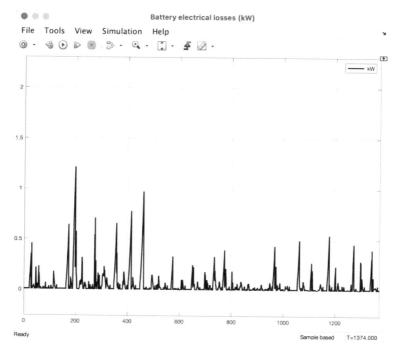

Figure 7.78: Battery losses during a 0° slope.

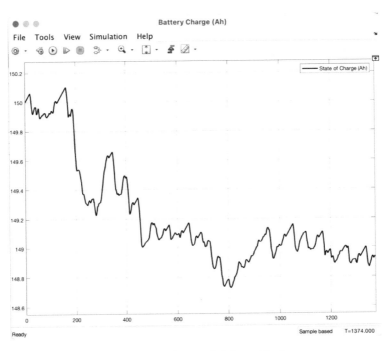

Figure 7.79: The battery state of charge in the model for a 0° slope.

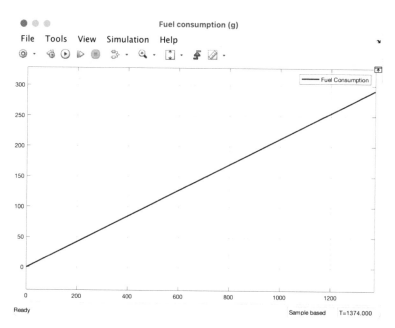

Figure 7.80: Engine fuel consumption at 0° slope.

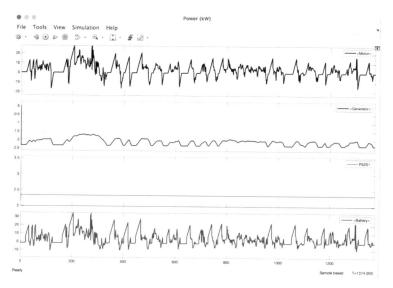

Figure 7.81: Engine, battery, generator, and motor power expended at 2° slope.

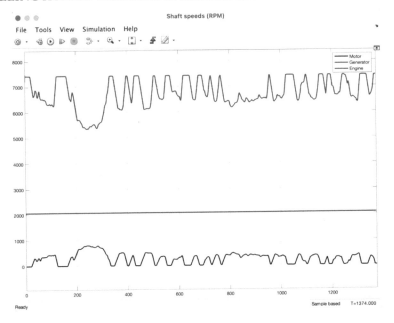

Figure 7.82: Speed comparison of the three sources at 2° slope.

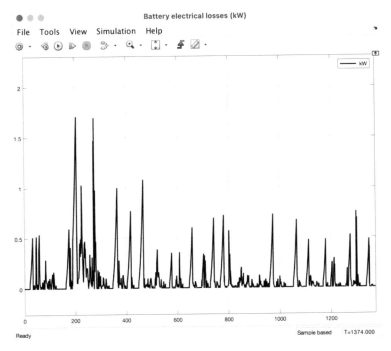

Figure 7.83: Battery losses during a 2° slope.

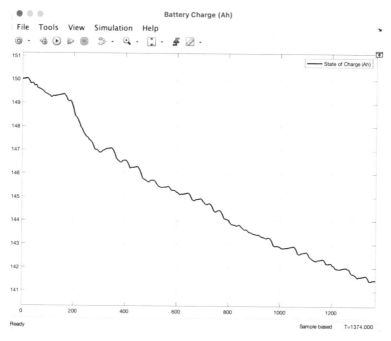

Figure 7.84: The battery state of charge in the model for a 2° slope.

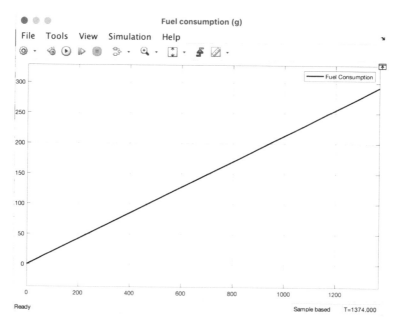

Figure 7.85: Engine fuel consumption at 2° slope.

7.4   OVERALL VEHICLE CONTROL

So far we have discussed models for three vehicle powertrain architectures for HEVs: the series, the simple parallel, and the power split architecture. The models were highly simplified to capture some of the essence of these architectures; to keep the models manageable, a lot of details, particularly details of many sub-systems and components, were not included. We have discussed before that HEV driving is made of different modes such as: battery alone, combined power, engine alone, power split, stationary charging, regenerative mode, etc. In a real system, these different modes are implemented through a complex control algorithm that helps keep the operation optimized at all stages and switch from one mode to another as driving conditions change. Demonstrating all of that was beyond the scope of this text. However, those who have mastered the basics can refer to this YouTube video: https://www.youtube.com/watch?v=EdBlOPHjdFg for instructions to create such a model, as well as refer to this page on Matlab's site: https://www.mathworks.com/matlabcentral/fileexchange/28441-hybrid-electric-vehicle-model-in-simulink to get access to such a model.

In this section we will explain how this control algorithm works using the example available on Matlab's site. Figure 7.86 shows an overall view of the entire model. This figure shows several subsystems, such as the Vehicle Dynamics, Electrical Subsystem, Power Split, Engine, and Control. The Vehicle Dynamics subsystem is similar to the Vehicle Body subsystem used in earlier examples in this chapter and models the different loads on the vehicle. The electrical subsystem comprises of the motor and generator, and uses the same approach as in the earlier examples in this chapter. The engine model is likewise similar to our earlier examples, as is the powersplit subsystem that models a planetary gear train. The one part of this model that was not present in the earlier examples is the control subsystem. In the previous examples, we took a simplified approach by running the engine continuously at either a certain speed or at a certain throttle position, the generator demand was also set at a constant value. Those two settings, while acceptable in certain situations, do not reflect the entire operational spectrum of the vehicle operation. As was shown in Figure 7.5 the vehicle controller needs to optimize the vehicle performance by turning certain device on or off and run the device at a certain setting depending on what the vehicle is doing. For example, when the vehicle is starting engine is turned off and the motor uses battery power to start with the motor and generator in operation. During normal cruising, both the motor and the engine help with the driving process, and during power boost the engine and the motor both drive at higher power, and during regeneration or braking mode the engine and motor are shut off while the generator output charges the battery. This model available on Matlab's site shows the details of a control algorithm that makes this type of switching between different operation modes happen. Here we have attempted to point out how this control algorithm works.

Figure 7.87 shows the Control subsystem. At the core of this control algorithm is the Model Logic subsystem, a state machine or a Stateflow (a Matlab tool) logic controller, that switches from one mode to another depending on driving conditions. The output of this state

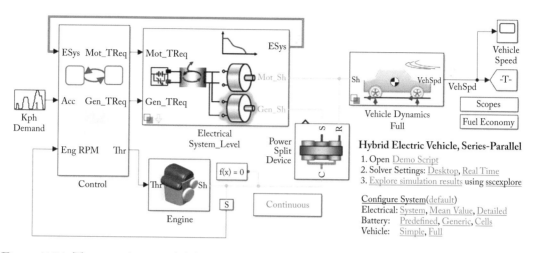

Figure 7.86: The complete model from Matlab's site.

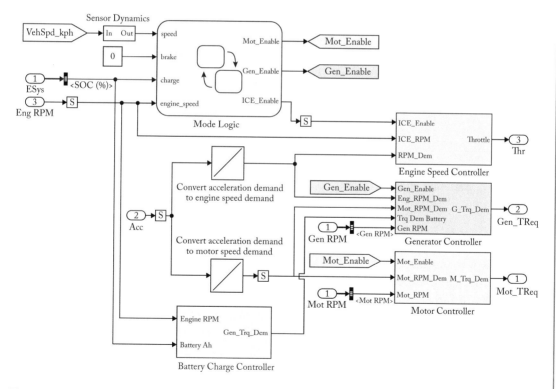

Figure 7.87: Control subsystem.

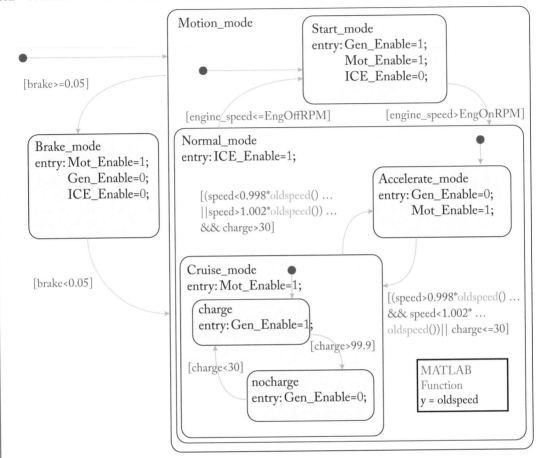

Figure 7.88: Stateflow decision logic.

machine is fed into three subsystems dedicated to control the engine, the generator, and the motor. There is another subsystem that monitors and controls the battery charge. The engine control subsystem (Figure 7.91) is a PI controller at its core and the output of this controller controls the throttle position for the engine. The generator subsystem (Figure 7.89) is also a PI controller and its output is a torque demand from the engine. The motor control subsystem (Figure 7.90) is similarly a PI controller that determines the motor torque demand. The Stateflow logic shown in Figure 7.88 shows how the switching between modes happen, such as the start mode and the normal mode, and within the normal mode there is the accelerate mode, and cruise mode. Also, there is the battery mode. For example, in the start mode the engine is off and the motor and generator are on. In the battery only mode the motor is on but the generator and engine are off, etc.

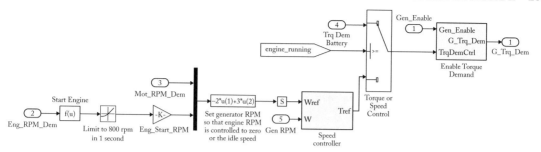

Figure 7.89: Generator speed controller.

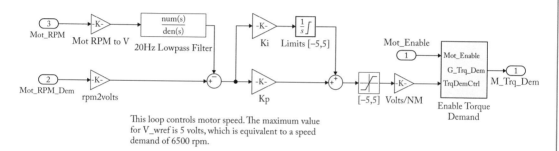

This loop controls motor speed. The maximum value
for V_wref is 5 volts, which is equivalent to a speed
demand of 6500 rpm.

Figure 7.90: Motor speed controller.

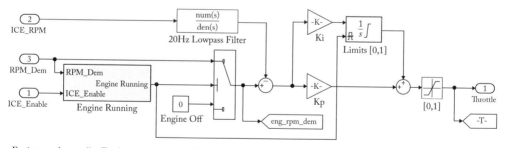

Engine speed controller. For demands below the idle speed
of 800 rpm, the speed demand is set to zero.

Figure 7.91: Engine speed controller.

After trying out the earlier examples from this chapter, readers are strongly encouraged to
use this example from the Matlab website to test out different parts of the control algorithm and
spend some time running this model and exploring in detail how different driving modes are
simulated. We are not including graphs and results from this model since the reader can easily
access this model from the link given earlier.

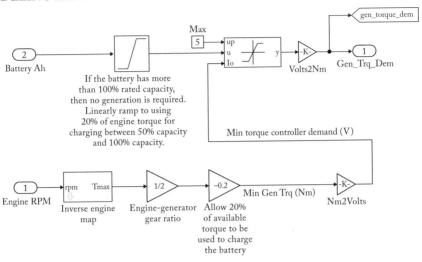

Figure 7.92: Battery charge controller.

7.5   SUMMARY

In this chapter we presented models for different vehicle architectures for HEVs. In the three examples provided we used simplified versions of power electronics and motors to focus more on the system model. Even though some of those models (detailed motor models and power electronics) were discussed in earlier chapters, we chose to use simplified versions in the architecture models here to avoid some of the complexities and hence keep our focus on the big picture. Therefore, d-q controllers and other control mechanisms didn't need to be included in the examples discussed in this chapter. As it should be obvious by now, Simscape modeling platform is designed such that models can be very modular. This means that a simplified model for a particular subsystem can be easily swapped with a more complex model of the subsystem. This same thing can be attempted with the models presented here as well.

In Section 7.4 of this chapter we discussed briefly a model that is available on Matlab's site where a complex control logic is used to simulate the HEV's entire operational spectrum including starting, speeding up, cruising, and braking modes. The controller is set up to ensure the right combination of power sources are used to optimize the power used for vehicle operation.

Author's Biography

SHUVRA DAS

Shuvra Das started working at the University of Detroit Mercy in January 1994 and is currently a Professor of Mechanical Engineering. Over this time, he has served in a variety of administrative roles such as Mechanical Engineering Department Chair, Associate Dean for Research and Outreach, and Director of International Programs in the college of Engineering and Science. He has an undergraduate degree in Mechanical Engineering from Indian Institute of Technology, and a Master's and Ph.D. in Engineering Mechanics from Iowa State University. He was a post-doctoral researcher at University of Notre Dame and worked in industry for several years prior to joining Detroit Mercy.

Dr. Das has taught a variety of courses ranging from freshmen to advanced graduate-level, such as Mechanics of Materials, Introductory and Advanced Finite Element Method, Engineering Design, Introduction to Mechatronics, Mechatronic Modeling and Simulation, Mathematics for Engineers, Electric Drives, and Electromechanical Energy Conversion. He led the effort in the college to start several successful programs: an undergraduate major in Robotics and Mechatronic Systems Engineering, a graduate certificate in Advanced Electric Vehicles, and thriving partnerships for student exchange with several universities in China.

Dr. Das received many awards for teaching and research at Detroit Mercy as well as from organizations outside the university. His areas of research interest are modeling and simulation of multi-disciplinary engineering problems, engineering education, and curriculum reform. He has worked in areas ranging from mechatronics system simulation to multi-physics process simulation using CAE tools such as Finite Elements and Boundary Elements. He has authored or co-authored four published books on these topics.

Printed in the United States
by Baker & Taylor Publisher Services